Methods for the Estimation
of Production of Aquatic Animals

Methods for the Estimation of Production of Aquatic Animals

Edited by

G. G. WINBERG

Zoological Institute
Academy of Sciences of the USSR

Translated from the Russian by

ANNIE DUNCAN

Royal Holloway College
University of London

1971

ACADEMIC PRESS
LONDON AND NEW YORK

ACADEMIC PRESS INC. (LONDON) LTD
Berkeley Square House
Berkeley Square
London, W1X 6BA

US Edition published by
ACADEMIC PRESS INC.
111 Fifth Avenue
New York, New York 10003

Library of Congress Catalog Card Number: 70-153528
ISBN: 0-12-758350-5

PRINTED IN GREAT BRITAIN BY
T. AND A. CONSTABLE LTD
HOPETOUN STREET, EDINBURGH

PREFACE

From the very beginning, one of the main aims of the International Biological Programme was to investigate all stages of biological productivity of eco-systems. This task has met with many difficulties, each habitat requiring the cooperation of various specialists who sometimes misunderstand each other. There are numerous gaps and weaknesses in our information and in our methods of research. It was our aim that the handbook in its original Russian edition should attempt to remedy this situation, particularly in relation to production estimates for both marine and freshwater invertebrates.

The Soviet edition was published in 1968 by the Minsk Higher School Publishing House under the auspices of the Soviet National Committee of the International Biological Programme of the Academy of Sciences of the USSR and the Belorussian Lenin State University, as a result of a decision taken at a symposium convened by the Soviet National Committee of the IBP and held at Minsk between 30th August and 4th September 1966. Forty-five Soviet biologists from twenty-one different scientific organizations, University departments, Institutes of the Academies of Sciences and State Research Institutes attended this symposium and contributed to its discussions. There it was agreed that Part 1 of the handbook, entitled " Methods for the Estimation of Production of Aquatic Animals ", should consist of instructions on the following topics: methods for expressing biomass; questions relating to the growth of aquatic animals, their duration of development and fecundity, particularly in relation to temperature and food; methods of determining the production of aquatic invertebrates with and without continuous recruit-ment of young; methods of estimating the production of communities and trophic levels. These methods and procedures were to be illustrated by examples from Soviet investigations already published or still in progress. The second part of the handbook was to consist of original papers, published under their authors' names, on various groups of animals, which were con-tributed to the symposium at Minsk in 1966.

An editorial board was set up consisting of G. G. Winberg (Chairman), G. A. Pechen' (Secretary), V. N. Greze, V. P. Lyakhnovich, B. M. Mednikov, A. P. Ostapenya and E. A. Yablonskaya. The text of Part 1 was prepared by the following authors: *Chapter 1*—G. G. Winberg; *Chapter 2*—A. P. Ostapenya with L. L. Lebedeva (2.2.1) and A. P. Pavlyutin (2.2.2 and 2.2.3); *Chapter 3*—G. G. Winberg (3.1 and 3.2) and G. A. Pechen' (3.3); *Chapter 4*—E. A. Yablonskaya, M. Y. Bekman and G. G. Winberg; *Chapter 5*—G. A. Pechen' (5.1, 5.2.1 and 5.6), V. N. Greze (5.2.2), E. A. Shushkina (5.3), G. A.

Galkovskaya (5.4) and G. G.Winberg (5.5); *Chapter 6*—V.P.Lyakhnovich and
V. V. Menshutkin (6.5). Contributions made during the symposium were also
used; for example, by A. F. Timokhina (4.4), B. M. Mednikov (3.2) and E. A.
Yablonskaya (6.2). Research workers from the Department of Invertebrate
Zoology and Experimental Biology of the Belorussian University helped
with the methodological section of the text.

This handbook therefore represented a first attempt to collate and
summarize the methods in general use in the Soviet Union for determining
the production of animals. When published in Minsk, it was not only one
of the first manuals to be produced by a National IBP Committee but was
also one of the first methodological handbooks on secondary production and
dealing with invertebrate animals. This reflected the early development of
production biology in the Soviet Union and it was hoped that it would prove
a useful stimulus for the development of fundamental research into the
productivity of water bodies.

It is therefore with pleasure that I welcome the appearance of an English
edition of Part 1 of the original handbook. It represents a useful summary
up to 1968, contributed by active workers in the field of production biology,
of methods generally employed in the Soviet Union together with their
theoretical basis, as well as a review of results of Soviet investigations carried
out before that date. These now routinely employed procedures and
results from quite early production studies (for example of E. V. Borutski
in the 1920's) are still not generally known outside the Soviet Union. It is
therefore hoped that the English edition will not only serve to disseminate
information about Soviet production biology but also that the theoretical
section on growth and metabolism, together with the methods based upon
this, will help to stimulate fundamental studies into the conception of trophic
levels of energy utilization. Despite the many difficulties involved, it is
apparent that quantitative investigations on energy flow in whole ecosystems
are vitally necessary. What is now needed is to improve our methods in
order to raise the quality of our results so that we may clarify the characteristics common to biological budgets and those which typify water bodies of
particular kinds. Now is the time to apply the concept of bio-energetic
budgets to elucidate the influence of organisms belonging to one trophic
level on the composition and biological characteristics of the preceding level;
that is, we need to examine those inter-trophic relationships which run in the
opposite direction from that of energy flow and, concurrently with investigations on the abiotic factors affecting the specific composition and biological
characteristics of the constituent organisms of each trophic level, to try to
further the development of the theoretical basis of the ecosystem concept.

The study of the productivity of whole ecosystems only becomes possible theoretically when based upon the concept of trophic levels of energy utilization, which enables all stages of the production process to be examined quantitatively. In fact, only relatively simple ecosystems have been examined in this way and it is still not very clear how useful is this concept of trophic levels for the study of biological productivity of more complex and larger water bodies. However, a new step in this direction was taken at the recent UNESCO–IBP symposium on the Productivity of Freshwater, held at Kazimierz Dolny, Poland, during May 1970. There it was possible to report on results of some Soviet IBP investigations more recent than those included in the 1968 handbook.

In the Soviet Union, as elsewhere, limnological investigations carried out by various groups inevitably differ in many respects, but some attempt has been made to present results from several lakes in a comparable form. The scheme adopted was that employed during the study of the biological energy budget of Lake Drivyata, reported by Winberg to the Third General Assembly of the IBP at Varna (IBP News no. 12, 22-24). The PF (Productivity Freshwater) Section of the Soviet National Committee of the IBP presented at Kazimierz the first results of studies on the most important trophic levels in bodies of water of various types and from different geographical regions. These water bodies ranged from the Karelian Lakes Krivoe and Krasnoe near the Polar Circle (M. B. Ivanova, Zoological Institute of the Academy of Sciences of the USSR), Lake Krasnoe near Leningrad (V. G. Drabkova, Limnological Laboratory of the Academy of Sciences of the USSR), three Belorussian lakes, Naroch, Miastro and Batorin (T. M. Mikheeva, Belorussian State University), Lake Drivyati (G. G. Winberg), the Rybinsk reservoir on the Volga River (Ju.I. Sorokin, Institute of Biology of Inland Waters, Academy of Sciences of the USSR) and the Kiev reservoir on the Dniepr River (D. Z. Gak, Institute of Hydrobiology of the Academy of Sciences of the Ukranian SSR). These lakes therefore range from the Polar Circle to the latitude of Kiev and differ considerably in size, depth and in the development of a thermocline. Most of them have been studied hydrobiologically over long periods of time. The Kazimierz report on these nine lakes will be published in the proceedings of the symposium but must be considered only as a preliminary summary. However, it will form a useful addition of more up-to-date information to the results of the investigations which are described in the present handbook.

1970   G. G. WINBERG

TRANSLATOR'S NOTE

Until the publication of the original Russian edition of this book, one of the weaknesses of the International Biological Programme has been the lack of a handbook dealing with how to measure secondary productivity of aquatic invertebrates, although many of these are important food species for fish. In the Soviet Union, there have been investigations on aquatic biological productivity since the 1930's, so that any Russian methodological manual on how to estimate it is of interest, particularly when edited by Professor G. G. Winberg who has been studying the various aspects of biological productivity for most of his professional life. Moreover, among the animals considered in the book were such groups of organisms as the chironomids, rotifers and cladocerans, which are numerically and functionally important components of aquatic ecosystems and whose production for various reasons is somewhat difficult and complicated to estimate. Furthermore, there was in the handbook one chapter with a theoretical exposition by Professor Winberg on the relationship between various kinds of animal growth and metabolism and between temperature and developmental rate, both of which are questions central to production biology. Finally, the handbook provided a critical review of the considerable body of Soviet research, published during the 1950's and 1960's, on the secondary productivity of aquatic invertebrates about which relatively little is known. For these reasons it seemed worthwhile to translate and publish an English edition of the handbook during the lifetime of the IBP.

During the winter of 1968–69, when I was at the Zoological Institute (Academy of Sciences of the USSR), Leningrad, Dr. M. B. Ivanova and I produced the first version of the translation which was seen by Professor Winberg. Subsequently I heard that the Translation Service of the Fisheries Research Board of Canada had been requested to translate chapters of the book, although not for publication; copies of these translations were very generously sent to me, for which I was most grateful. The Canadian Translation Service has considerable experience in translating Russian biological literature which I was glad to draw upon, particularly for identification of the Russian common names of animals. For all other changes and for the style of the language I bear full responsibility.

Certain additions have been made. An index has been provided for the English edition which adds to the ease of finding one's way about the book. To the Russian references, which have been checked as far as was possible in England, English titles have been given as well as a reference to any existing translation. This was the work of Miss M. J. Burgis and involved a great deal of painstaking labour but it has added considerably to the usefulness of the translated edition in both its rôles: as a manual on secondary productivity and as a review of Russian research. Russian names have been transliterated according to the International System for Transliteration of Slavic Cyrillic Characters and apologies are extended to those biologists whose names are given in a form which they do not like. It was decided, together with Professor Winberg, that Part 2 of the handbook, which consists of original contributions by various authors, should not be included in the British edition.

June 1971   ANNIE DUNCAN

CONTENTS

Chapter 1

INTRODUCTION

1.1 General Aims in Estimating Production of Aquatic Animals

Rational exploitation of biological resources is hindered by the inadequately developed state of theoretical ideas on biological productivity of ecological systems. Consequently, development of a theory of biological productivity is one of the central aims of contemporary biology, which requires the organized and cooperative efforts of specialists in many fields of biology from several different countries. Recognition of this problem forms the main theme of the International Biological Programme which has initiated a programme on the study of the fundamental laws governing the biological productivity of terrestrial and aquatic communities, the understanding of which is needed for a better exploitation of natural biological resources in the service of man.

The general direction of research on biological productivity in fresh waters, in line with the aims of the International Biological Programme, was considered at a special symposium held in November 1965 and convened by the Soviet National Committee of the International Biological Programme under the auspices of the Academy of Sciences of the USSR.

The "General principles of the Soviet National Programme of research into the productivity of freshwater communities" (1966) were formulated at this symposium. In this publication it is stated that "in its theoretical aspect, the productivity of freshwater communities is considered to be a single, multiphased process, all of whose stages are interrelated and conditioned by natural laws".

Products utilized by man belong to the various different stages of the production process. Where a useful product, such as fish, is far removed from the original sources of matter and energy of the production process, experience has produced all kinds of techniques for increasing fish productivity in bodies of water. Such interventions can be used not only to control the primary links in food chains or the primary phases of the production process (by mineralization of waters and so on), but subsequent intermediate stages (by introduction and acclimatization of food organisms and other methods of improving the food supply), or even the final stages of production (by fish management or other measures for improving fishery economy).

The process of production in aquatic ecosystems acts by means of trophic

B   1

interrelationships of organisms which result in the transfer of quantities of matter and energy from one trophic level to subsequent ones.

Consequently, it is fundamentally important to know those characteristics of individual organisms which determine their interaction with organisms from adjacent trophic levels, for instance their food requirements. In studying the production process, the biomass, production and individual functions of organisms from different trophic levels should be expressed in comparable units. Such units may be units of mass and energy.

The expression "units of mass and energy" refers here to milligrammes, grammes, or other weight units for dry or wet organic matter, organic carbon, organic nitrogen, and to calories, i.e. units which are interchangeable with the help of certain coefficients and with a greater or lesser degree of approximation.

In each individual case, in order that the biomass and, especially, the production of a species-population or communities may be expressed finally in calories, it is necessary first to have detailed information about the abundance, the individual weight, calorific content and many other specific characteristics of the organisms being studied. In this way, expressing biological data in units of mass and energy supplements rather than alters the results of a careful investigation of the specific characteristics of individual species or other biological features. It arises from the nature of these results and makes high demands upon their quality.

In organizing and carrying out work in a study on the productivity of fresh waters, it is essential, in the first place, that the collection of samples is arranged so that they are representative of the whole water body and, in the second place, that the final results of the investigation can be presented in such quantities that may be related to the body of water as a whole.

The first requirement can be accomplished by locating rationally the sampling stations at which observations are made and the depths at which samples are taken. In order to do this some preliminary information is needed about the characteristic zones existing in the body of water, their relative dimensions and the main distributional patterns of the organisms inhabiting them.

The second requirement will be satisfied by suitable handling of the original data. In particular, the calculation of mean values must be weighted by the relative volume of the water layers involved, etc. Only data obtained this way can be used to compute values (indices) which apply to the entire water body.

The need for such calculations does not imply that final results must be expressed as the biomass or production for the whole water body. This not infrequent practice must be decisively rejected as the very large figures involved for the total biomass or total production of a whole body of water are difficult to assimilate and useless for comparative purposes. Thus, after such values are calculated for the whole water body, they must then be converted to a unit surface area and expressed as kilogrammes per hectare or grammes and calories per square metre etc., that is, converted to a form that can be compared with similar results from other water bodies.

Biomass, that is the weight of a population of animals in a particular area of a biotope, zone or entire body of water at a given moment, is the resultant of the opposing processes of reproduction and growth on the one hand and elimination on the other.

Biomass evaluates the degree of quantitative development of species inhabiting a body of water. However, from biomass alone, nothing can be said about the production processes of a species, namely the intensity of growth and reproduction of the organisms in a species-population. It is absolutely essential to get to know these population characteristics if we are not only to describe the qualitative or specific characteristics of trophic chains in a particular water body but also to estimate quantitatively the individual trophic links and the transfer of matter and energy from one trophic level to the next.

During the last ten years a great number of new reservoirs have been built whose flora and fauna have developed under the influence not only of natural processes but also of man. Often these reservoirs have been stocked with young fish of valuable species with a view to commercial exploitation.

Huge sums are being invested in the construction of fish factories and hatchery farms in the lower reaches of the Volga, Don, Kura, Kuban and other rivers to compensate for the loss of natural spawning grounds of migratory and semi-migratory fish caused by the regulation of the flow of these rivers. The possible scale of artificial rearing of young valuable fish species is mainly determined by the abundance of the food supply (food base) at the nursery and fattening areas in the sea, reservoir and elsewhere, i.e. by the productive properties of the communities and populations of the food species.

Without information on the reproductive rate of the food supply of the growing fish, on the breeding intensity and on the growth of the invertebrates which make up the fish food organisms, it is not possible to predict the expected level of fish production from any particular water body or from a unit area of its surface. It is well known that at present, due to the poorly developed state of theoretical ideas on biological productivity, even detailed hydrobiological information can be applied only to a limited extent to a fishery in order to estimate its potential. Conclusions so frequently found in fishery publications as to whether the food supply is "under exploited" or is "fully exploited" are not based upon any quantitative assessment but to a considerable extent upon arbitrary and subjective opinions. Such a state of affairs is intolerable in matters of such great practical importance and must be changed by the development of theoretical ideas on biological productivity. We need methods which will enable us to compare quantitatively the various stages in the production process in different types of water bodies and will allow us to reject subjective guesses about the relationship between the food supply and the fish yield of a water body. Good methods of computing production of aquatic animals are important for solving these problems.

Although estimates of the production of aquatic animals are needed for solving problems in production hydrobiology, they will also prove useful in the more applied aspects of hydrobiology. Here our most immediate need is to clarify how aquatic organisms are involved in the self-purification of polluted waters. When considering these complex matters in a general way we must consider the part played by these organisms in the mineralization and destruction of unstable polluting substances as well as their stabilization by transformation into the living tissues of the organisms. The former process is basically proportional to metabolic intensity multiplied by biomass and the latter to growth or production. The process of self-purification is the function of a complex multi-specific community and is achieved to a large extent by the trophic relations between its species. These interactions can be quantified, and the normal patterns of the self-purification cycle clarified, only by computing the production of the species-populations involved.

It is well known that at present numerous studies are being carried out on aquatic primary production which, using modern methods, is measured and expressed in grammes of organic carbon per square metre or in calories per square metre for 24 hours or a year. In order to assess how and to what extent primary production is utilized by the subsequent stages of the production process, it is necessary to know not only the numbers and biomass of the organisms belonging to the intermediate trophic levels but also their production. It is therefore imperative that methods should be available for computing these values which can be applied to the species-populations of different kinds of aquatic organisms. For example, many authors have attempted to determine the production of planktonic and benthic species-populations, using various different methods and formulae. This data is scattered throughout the extensive hydrobiological literature and has not yet been collected together, nor have the various methods used for computing production been compared or their validity, accuracy and limits of applicability critically evaluated. The present work is the first attempt at such a comparison. The aim of this book is to help in selecting and applying the various methods available for computation of production, and to stimulate the development and improvement of other methods so that it will to some extent serve as a guide. It must be remembered, however, that the methods described here are still not generally used and therefore their suitability and whether they require improvement still need to be verified. They can never be applied mechanically but require a creative, critical approach with good understanding of the basic requirements for each method and a sound judgement regarding the limitations within which they may be applied.

As many of the methods for computing production can be used for both freshwater and marine animals, examples have been taken from the publications of authors working in marine or brackish water habitats.

Estimates of production only become realistic when based upon good quantitative data on the abundance and biomass of the species-populations inhabiting a body of water. However, in practice such data are difficult to obtain because our methods for estimating the density of aquatic populations are very imperfect. The question of whether the methods generally used in hydrobiology to estimate the abundance of aquatic inhabitants provide good data is a difficult matter and requires separate treatment, which is not contained in this book. In the following chapters it has been assumed that realistic data on numbers and biomass of a particular population can be obtained with an adequate degree of accuracy and reliability.

Similarly, no examination of the statistical methods for analysing the quantitative data has been included in this book as these are available in the appropriate textbooks. All estimates of production based on mean values of biomass and numbers of individual species have a statistical nature and at every stage in the computation these quantities must be tested for reliability by the usual methods of biological statistics.

Most attention is directed towards computing the production of individual species-populations which at the present stage is entirely justified. But this at once raises the question of how to relate the production of the individual species-population to that of the ecosystem as a whole (this question is touched upon in Chapter 6). It is unfortunate that even indisputable theoretical ideas are not always observed in practice. For example, even now summed values of zooplanktonic or zoobenthic productions are presented, although these consist of both predatory and non-predatory forms. It is quite obvious that such a summing of productions of species-populations belonging to different trophic levels is nonsensical and incorrect.

We are only just beginning to realize how to make quantitative assessments of the contribution from the production of one species-population to one or other of the final products, of "use" or "useless" to man or rather what is beginning to become clear is how to approach such an investigation. There is no doubt that one approach is the energetic treatment of the biological cycle by a study of the "flow of energy" within the ecosystem. This is generally accepted in contemporary ecology and hydrobiology. At the same time it is becoming increasingly clear that any energetic treatment of the various stages in the biological cycle must be accompanied by studies of population dynamics, age structure, mortality and reproductive rates, etc., that is, by the concepts of population ecology. This will be illustrated by many examples in the following chapters, although the subject-matter of this book is strictly limited to methods of estimating the production of aquatic animals, and other extremely important questions in present-day ecology and hydrobiology have been omitted, even though many of them are relevant to production methods. This is especially true of the ecological–physiological methods for determining food rations, metabolic intensity, food assimilability and others which are not included here, although the results of these determinations are needed for some production computations, as described later. These matters deserve separate treatment.

Thus, while dealing with methods for estimating the production of aquatic animals, we encounter many unresolved and unclarified problems of hydrobiology, the very existence of which should stimulate their investigation. The need for such quantities as growth rates, metabolic intensity and reproductive or mortality rates for estimates of production should encourage more experimental ecological and physiological investigations and field hydrobiological research.

It is to be hoped that, with more accurate methods and with greater precision in the theoretical ideas of biological productivity, the results of hydrobiological investigations will be applied more effectively to problems of practical importance in fisheries, sewage disposal plants and the water supply industry. For instance, it should no longer be necessary for a fishery to have to decide on the degree of food utilization in a particular water body on the basis of vague and subjective ideas. This is a direct consequence of poorly developed theoretical concepts. Every further development towards a theory of the biological productivity of water brings us nearer to our final aim of regulating the productivity of waters.

1.2 Terms and Symbols

Concepts and values used in determinations of animal production have been understood and designated in different ways by different authors. As this problem stands at present it is not easy to achieve general accordance about the terms and notations to be used for denoting the quantities involved. Nevertheless, as data from the work of many authors have been cited in the following chapters, it was necessary to adopt some sort of uniform system of terms with their corresponding literal symbols. Applying the criterion of common usage, those symbols adopted by the majority of scientists have been used whenever possible.

Firstly, it is necessary to clarify in what sense the term "production" is used. In agreement with general trends in recent years, the term "production" is held to have the same meaning as "net production" which is used by some authors, mainly those writing in English (Clarke, 1946); these authors consider it necessary to distinguish between "net production" and "gross production" and they understand that "gross production" includes not only the growth increment but also the metabolic loss for the same interval of time.

The meaning of the term "production" which has been adopted recently originates from the well-known definition of Thienemann (1931). The production of a species-population for a known period of time is considered to be the sum total of growth increments of all the individuals existing at the start of the investigated period and remaining to the end, as well as the growth of newly born individuals and those which, for various reasons (due to being consumed, dying or other causes), do not survive to form part of the final population biomass existing at the end of the period. As in studies of individual growth, the growth increment is defined as the increase in the amount of living or organic substance of the species in question as well as the energy included in it (Brotskaya and Zenkevich, 1936; Ivlev, 1945).

From this widely accepted point of view the term "gross production" becomes superfluous. It is entirely equivalent to the commonly used term "assimilated food", the energy and substance of which, as we know, goes into growth or production as well as into the energy requirements of metabolism (expenditure on metabolism). Similarly, there is no need to attach the adjective "actual" to the term "production". In certain cases when we are dealing not with the actual but the so-called potential production, this latter term can be used in its present form. Potential production is the calculated value of production under ideal conditions in the absence of any kind of limitation on growth or reproduction other than natural mortality and, as such, gives some idea about the productive potential of the species (Embody, 1912; Mordukhai-Boltovskoi, 1949; Briskina, 1950; Pidgaiko, 1965). In some cases it may prove valuable to compare this quantity with the production actually achieved (Bekman, 1954). Some authors (Nelson and Scott, 1962) have used the term "apparent production" for characterizing the quantities resulting from summing all the biomass increments observed during a particular period of time. For example, if the biomass curve for the period in question has three peaks, the three quantities representing the differences between the maximum biomass and the preceding minimum are added together. The precision of such estimates of production depends on the prevailing conditions and the specific characteristics of the system. That is why the expression "apparent production" has no clearly defined meaning and why, whenever such a term is needed, it would be better and more accurate to use "biomass increment" or "the sum total of biomass increments for a particular period of time".

Recently the term "productivity" has been used frequently as a synonym for "production", especially by authors writing in English (MacFadyen, 1965). In the works of Soviet authors the term "productivity" has come to mean the characteristic property of a particular population, community or water body which is responsible for the large or small level of production. We should recognize this use of the term as the correct one as well as the need for distinguishing between the terms "productivity" and "production".

As the growth of living matter is the basis of production, estimates of it are expressed in units proportional to mass (wet weight, dry weight, the weight of organic carbon and so on) or in the equivalent units of energy (calories) and related to time (24 hours, a year, etc.).

In certain cases, either for special purposes or as an intermediate stage in the calculation, production may also be expressed as the number of individuals born in a unit of time.

Like biomass, abundance and other characteristics of a species-population under particular conditions, production may be related to a community, an ecosystem or an entire body of water or to units of volume (litre, cubic metre, etc.) or to units of surface area (square metre, hectare and so on).

The most meaningful quantities are those calculated for an entire ecosystem and then converted to units of surface area. In this way the production of diverse ecological systems can be compared.

Because secondary intermediate production is investigated together with primary production and the final stage of production, i.e. the fish yield expressed in kilogrammes per hectare, it is necessary to convert biomass, abundance and production to surface units. Often in studies on phyto- and zooplankton it is only by expressing biomass, abundance and production in both volume and surface units that any conclusions can be reached about biological production or that the normal patterns of the process of production can be clarified.

In the following chapters, the Latin alphabet has been adopted to denote the most important values met in calculating production.

N (*sometimes* n)—*number of individuals*. As in the case of all subsequent symbols that refer to a population, N can be related either to an ecosystem or a water body as a whole or to volume units (litre, cubic metre, etc.). In the latter case N may be regarded as an index of the density of animals. N is usually qualified by a subscript (N_0, N_t, N_1, etc.) in order to define more clearly whether the initial, final or intermediate numbers are being cited.

In addition the letter N, without a subscript, is one of the constants in growth equations (see Chapter 3).

w—weight (mass) of one individual. This quantity can be expressed in units of wet or dry weight, of organic carbon, etc., or in their corresponding energy units (calories). When it is necessary to emphasize that we are dealing with mean weight, then, as in all other similar cases, a bar is placed over the letter—$\bar{w}$.

W—maximum (definitive) weight (mass) of an individual, expressed in the same units as w.

B—biomass. The biomass indicates the total weight (mass) of all individuals of a population or the part being studied. Thus $B = \sum_{1}^{n} w$ or $B = N\bar{w}$. It is of course expressed in the same units as w. Sometimes the term "biomass" is used with reference to one individual, and then it replaces, unnecessarily, more precise terms such as "the dry weight of an individual" or "the calorific equivalent of the weight of an individual". This is undesirable and the word "biomass" should be used only with reference to populations or communities.

t—time. Thus, N_t is the number of individuals at time t, w_t the weight at time t, and so on. However, in other contexts t can refer to temperature.

D—the duration of development. The stage of development to which the symbol refers is indicated by a subscript, for example D_k is the duration of copepodite development.

$D^{-1} = 1/D$—*the rate of development,* representing the fraction of the total development which occurs during one unit of time.

Q—total metabolism of an individual, expressed in units of volume or weight of oxygen consumed or as the quantity of energy released (calories, etc.) per unit time at a known temperature. Winberg (1956) defines the various types of metabolism generally recognized in Soviet literature.

Q/w—rate of metabolism, expressed in the same units as Q but related to a unit of weight.

T—metabolic loss, expressed in the same units as w. In principle T represents the loss in body weight of an organism when it has no food. Metabolic loss may be represented either in absolute values when it refers to one or a number of individuals or in relative values when it is related to w. In the latter case, by analogy with Q/w, it should be denoted as T/w and expressed as fractions of w per unit time; that is, this quantity has the dimension of t^{-1}.

P—production of a population and the growth increment of one individual. In the works of Winberg and other Soviet authors this value is designated by the Russian character п. Calculated values of production are expressed in the same units as w and are related to time.

R—ration, the quantity of food ingested per unit time by an individual or by a whole population (it is designated by the Russian character P). R can be compared with P and T when it is expressed in similar units.

A—the assimilated part of the ration. It is clear that $A = P+T$.

P, R, A as well as T can be expressed either in absolute or relative units which have the dimension of t^{-1}. In the former case this is the quantity of matter or energy associated with one individual or with a known number of individuals (a population) per unit time. In the latter case it represents a fraction of w, if referring to an individual, or a fraction of B, if referring to a population, per unit time.

Perhaps it might be expedient to distinguish relative magnitudes not by capital letters but by small ones (t, p, r, a). However, this is not the usual practice in publications. Consequently capital letters only are used for both conditions in the following chapters. It will be clear from the context which concept is meant in each case.

E—elimination, namely the portion of the population production which is eaten or removed in some way. It is expressed in the same unit as P.

1/U—assimilability, a non-dimensional coefficient, representing the ratio and its assimilated part.

$$1/U = A/R, \text{ consequently } (1/U)R = A.$$

K_1—*the coefficient of consumed food utilized for growth* (Ivlev's first order coefficient of food utilization and the coefficient of ecological efficiency of Odum and other authors). This non-dimensional coefficient equals the ratio of growth increment of an individual or of a population production to its ration, $K_1 = P/R$.

K_2—*the coefficient of utilization of assimilated food for growth* (Ivlev's second order coefficient and Odum's efficiency of tissue growth), $K_2 = P/A$.

Other coefficients and constants in subsequent chapters are explained in the appropriate sections of the text.

Chapter 2

BIOMASS AND HOW TO EXPRESS IT

2.1 Determination of Biomass

By biomass we mean the mass of organisms in a body of water or per unit of its surface area or volume. It can be expressed in units of wet or dry weight and also in units proportional to these (carbon or nitrogen content or the quantity of oxygen required to oxidize the organic matter), or in units of energy.

It is not recommended that the term biomass be used to refer to a single individual when speaking of its weight or quantities proportional to weight.

Some authors, especially in the fishery literature, use the term "residual biomass", without any precise definition of its meaning, usually in the same sense as the term "biomass". Use of this superfluous term is not recommended.

2.2 Methods of Determining Wet and Dry Weight of Aquatic Organisms and their Ash Content

2.2.1 *Wet Weight*

In addition to the direct weighing of aquatic organisms, a variety of indirect methods are used for determining wet weight. Most of these indirect methods involve determining the body volume of the animal. The specific gravity of the wet matter of the body of aquatic organisms is usually considered to be equal to unity. The majority of indirect methods have been developed for situations where it is necessary to determine the wet weight of small organisms, especially planktonic organisms.

One of these methods is that of computing a geometric equivalent. In this method the form of the body is approximated to some kind of simple geometric shape (sphere, ellipse, cylinder, cone, etc.). Measurements are made under the microscope for calculating its parameters and its volume is then computed from standard formulae. The method of Lohman (1908) may also be employed in which models of the organisms being studied are used to compute their volume. A model may be made from Plasticine on wire axes, the lengths of which are proportional to the measurements made on the object being studied.

Methods of determining individual wet weight from the volume of the organism being studied have a number of inadequacies common to them all,

11

the principal inadequacies being the considerable labour required and the poor accuracy obtained, especially when the body of the organism has a complex form.

In addition to the computational methods of determining volume, apparatuses of varied construction are frequently used for measuring the volume of the organisms in a sample, i.e. volumometers; for example, the volumometer of Usachev as modified by Greze (1948). In work with volumometers of this type the change in level of the liquid in the apparatus which occurs after the introduction of the organisms being studied is measured, and from this their volume is computed.

Volumometric methods have been used mainly to determine the biomass of plankton and then give useful results with sufficiently large samples (of the order of several hundred milligrammes).

For determining the wet weight of comparatively large objects or when working with samples weighing several milligrammes or more, direct weighing is used. The fundamental difficulty encountered in direct weighing is that, before weighing, the organisms must be freed of external water. In the drying that follows immediately after the removal of the external water, water begins to be extracted from the composition of the body. It is very difficult to catch the moment when the organism is free from external water and has not yet begun to lose internal water. Rezvoi and Yalynskaya (1960) describe a method for avoiding the preliminary drying. They recommend the following procedure.

"First of all evaporate the surrounding water from the organisms contained in a weighing dish. While this is being done no evaporation of internal water occurs, because of protection by the outer coverings of the body. Checking the weight continuously we observe a comparatively rapid decrease before all the external water has been evaporated. There follows a plateau, or a well-marked retardation of loss in weight. Later weight again begins to decrease but at a slower rate, because internal water is evaporating through the external body covering. By measuring the weight at unit intervals of time we may construct a curve of weight loss. The moment at which the surface water disappears will be marked either by a complete standstill, that is a horizontal segment of the curve, or by a bend in the curve beyond which the slope increases. The weight corresponding to the plateau or inflection will be the original weight of the organism."

This method does not, however, always give accurate results. For example, Sushchenya and Vetrova (1957) have shown that curves of weight loss against time frequently do not have any clearly developed inflection, and therefore it is not clear which point corresponds to the termination of the evaporation of external water. Analogous results were obtained by Lebedeva in working on certain planktonic organisms, and by Pavlyutin in work with chironomids (unpublished data). Thus the method of Rezvoi and Yalynskaya cannot be recommended for determining wet weight of water organisms.

For determining the quantity of external water methods have been proposed in which previously weighed animals are placed in a solution of a particular substance of known concentration. A certain quantity of water is introduced with the organism and the amount of water present is then estimated from the degree of dilution of the solution. For this purpose Gaevskaya (1938) used a solution of glucose and she has proposed another, more sensitive, modification of the method which is described below.

The organisms are placed in a solution of glucose of known concentration and are then transferred to a vessel containing a known quantity of water. In order to find the quantity of external water introduced with the animals, the quantity of glucose added to the water with the organisms is determined by the method of Hagedorn and Jensen and this is considered to be equal to the amount of external water. This method is rather difficult and time consuming and apparently for these reasons it is not widely used for practical hydro-biological investigations.

Most workers, while developing different variants of the direct method for determining wet weight, tried to standardize the procedure for drying the organisms. For example, Borutski (1934) recommended a one-minute period of drying before determining the wet weight of benthic organisms. Ulomski (1951) proposed drying a sample of planktonic organisms until no more wet spots appeared on the filter paper. Mordukhai-Boltovskoi (1954) recommended a similar drying method.

Tests on the various procedures available for determining the wet weight of planktonic organisms show that Ulomski's method gives relatively good results. It appears that this procedure can also be applied to benthic organisms. The author describes it as follows:

"The animals, transferred to a watch-glass and counted under a binocular microscope or magnifying glass, are sucked up in a pipette. The organisms, settled and concentrated at the tip of the pipette, are transferred with 3–4 drops of water onto a 2 cm square of No. 20–25 bolting gauze which rests on a ring lying on top of filter paper. For some time the water containing the animals neither wets the gauze nor is absorbed by the filter paper. By lifting with forceps one side of the square and holding down another side with a needle, the drop of water soaks into the filter paper but, in passing through the gauze, wets its underside only slightly and does not disperse. The wet spots on the gauze should have a uniform area when the same number of drops of water are used. The organisms usually settle evenly and in a single layer on the wet surface of the gauze. The gauze with the organisms is now placed on a dry filter paper and, by alternately lifting and lowering it with forceps and the needle, is dried until the paper no longer shows damp areas. At this point the gauze square with the organisms is placed in a weighing vessel, covered and weighed. The weighing is repeated. To do this the gauze is rinsed again in the watch glass and dried to a constant weight, the organisms having been counted and again placed onto the dry gauze with a pipette. The mean of the two weighings is then calculated."

By careful and accurate standardization of the procedure and drying time, it is possible to obtain results with an error of 5 per cent or less.

Determining the wet weight of molluscs, particularly Lamellibranchia, is difficult and liable to error due to the water contained in the mantle cavity. The water in the mantle cavity must be removed before the organism is weighed. The shell of small species can be punctured with a fine needle and the mantle cavity water sucked up with filter paper, after which the animal can be weighed.

Many hydrobiologists have to determine wet weight from fixed material. It is generally known that the wet weight of organisms fixed with formalin differs from their living weight. Moreover, Borutski's (1934) data show that the weight of fixed organisms changes during storage and stabilizes only after four months. In benthic animals the difference in weight between fixed and live organisms is about 2–4 per cent. Little is known about this problem in planktonic organisms.

Methods for determining wet weight require further development. For the present, Ulomski's method can be recommended for mass determination of wet weight. With fixed organisms the changes in wet weight due to fixation must be considered. More information is needed on the differences in the wet weight of fixed and fresh objects.

2.2.2 Dry Weight

By dry weight we mean the constant weight of the totally dehydrated body. Several methods of drying biological material to a constant weight are available. These include drying in a desiccator over various drying agents (i.e. $CaCl_2$, H_2SO_2, silica gel), under vacuum, by lyophilization, under infrared rays or in an oven.

When large samples of material have to be dried the time it takes to obtain a constant weight is not unimportant. To dry completely a sample weighing 15–20 mg in a desiccator takes 20–25 days (Lovegrove, 1962) but only a few hours in an oven. It is usual, therefore, in hydrobiology to determine dry weight by the simple and rapid method of drying the samples in an oven. Other methods are used less frequently and usually only when there is a risk of the lipids or other compounds being oxidized.

Temperatures ranging from 50 to 105°C have been recommended for drying biological samples. The constant weights obtained at different temperatures differ so slightly that they can be disregarded in very large samples. The unpublished data of Pavlyutin show that the constant weight of *Chironomus thummi* dried at 50°C differs by only 3·5 per cent from that dried at 105°C. The duration of drying, once constant weight is reached, does not affect the results, provided temperatures do not exceed 105°C.

Since dry weights obtained at temperatures ranging from 100 to 105°C differ little from those obtained at lower temperatures and drying at the higher temperatures takes much less time, temperatures in this range are recommended for large samples of water organisms. At 100–105°C samples of 100–300 mg wet weight reach a constant weight in 2–3 hours.

The balance used to determine dry weight should permit a weighing accuracy of not less than ±1–3 per cent.

2.2.3 *Determining Ash Content*

The ash content of a species may vary considerably and apparently depends both on its physiological state and ecological conditions. Table 2.1 shows the variation in ash content of the tissues of animals belonging to a given species.

TABLE 2.1

The ash content in the body tissues of some freshwater molluscs (Birger, 1961), in percentage dry weight.

Species	Mean ash content	±σ	Coefficient of variation
Unio tumidus	15·8	4·8	30·3
Monodacna colorata	11·3	3·3	29·1
Unio pictorum	12·8	5·9	42·2
Viviparus viviparus	30·3	6·0	19·8
Anodonta piscinalis	14·1	2·0	15·4
Dreissena polymorpha	17·5	2·7	15·4

The ash content in the bodies of aquatic organisms is often determined by incinerating samples in a muffle oven. The temperature at which the sample is incinerated is known to affect the result greatly. Some mineral salts present in the body of aquatic organisms decompose at certain temperatures, discharging gaseous products, resulting in a reduction in weight of the mineral fraction left after incineration.

According to Grove *et al.* (1961) significant losses in potassium and sodium occur at temperatures higher than 550°C. Between 400 and 450°C even prolonged incineration does not produce noticeable losses of these elements.

The main skeletal material of many invertebrates consists of $CaCO_3$. Paine (1964) has shown that this compound decomposed more effectively at temperatures greater than 550°C, with the production of CaO. This can decrease the weight of the mineral fraction by 44 per cent. The decomposition of mineral salts at their critical temperature occurs very abruptly and within a small temperature interval (Fig. 2.1).

It is necessary to bear in mind when making ash determinations that a considerable temperature gradient (up to 50°C) may exist inside the muffle oven. The crucibles should therefore be grouped compactly together in the centre of the oven, equally spaced from the front and back walls, so that the determinations will not be affected.

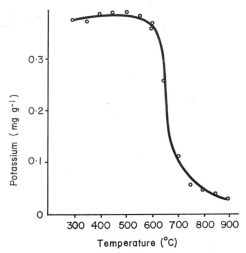

FIG. 2.1 The potassium content of ash incinerated at different temperatures in a muffle oven (Grove *et al.*, 1961).

Sometimes ash content is determined from the weight of mineral compounds remaining in a bomb calorimeter after combustion of the organic material of the sample at temperatures approaching 1000°C. Such ash determinations are always somewhat undervalued due to the thermal decomposition of some of the salts (the problems involved in this method of ash determination are examined in more detail in Section 2.3.1).

Considerable errors can arise in determining the ash content of aquatic animals with skeletons of $CaCO_3$ (for example, shelled molluscs) because of the water of crystallization. When samples are dried at 105°C the water of crystallization is retained but it is lost during incineration. It is recommended that samples should be incinerated in a muffle oven at temperatures not exceeding 500–550°C. For complete combustion of organic matter in a sample weighing about 100 mg incineration for between 20 and 24 hours is sufficient.

The determination of the ash content of molluscan tissues is best carried out on bodies freed from their shells.

2.3 The Calorific Value of Aquatic Organisms and Methods of Determining it

2.3.1 *Calorific Values of Actual Bodies*

To express weight of organisms in terms of units of energy their calorific value ("kaloriinosti") must be known. This is the energy content of a unit mass of material studied. Some hydrobiologists have also used the term "kaloriinosti" to denote either the number of calories contained in the body of a single organism, or the biomass, expressed in calories, and this has caused some confusion with the above definition. It is important to distinguish carefully between the energy value or energy equivalent of the total weight of an animal (in terms of calories per individual), the energy value or energy equivalent of the biomass (in terms of calories per cubic metre, calories per square metre, or calories per water body, etc.), and the calorific value ("kaloriinosti")—that is, the energy value of an organism related to one unit of weight (in terms of calories per gramme of wet substance, or of dry ashless material).

Accumulated information suggests that the calorific value of the dry body substance of water organisms varies between 0·2 and 8·0 kcal g^{-1}, mainly because of variations in the proportion of mineral to organic fractions. Relatively less significant are the differences in the chemical composition of the organic matter of the body in various organisms.

Ostapenya and Sergeev (1963) have demonstrated that the calorific value of dry matter in the most diverse organisms, marine, freshwater, planktonic and benthic, is directly related to the ratio between the organic and mineral fractions in the dry material (Fig. 2.2). This relationship is expressed by the equation

$$Y = 0·0559 \ X \ \text{at} \ \sigma Y/X - 0·28 \ \text{kcal g}^{-1}$$

where Y is the calorific value in kilocalories per gramme dry weight and X is the percentage organic matter present in the sample.

The square error (sample standard deviation from regression) calculated for the equation just given shows that 68 per cent of the data deviated from the regression by ±5 per cent. Therefore, for all practical purposes, once the ash content is known it is possible to calculate the calorific value of dry material from the equation with a sufficient degree of accuracy.

It follows from the equation that the calorific value of various aquatic organisms lies within the limits of 5·59 ±3σ, with a probability of 0·997, that is, between 4·74 and 6·42 kcal g^{-1} organic matter with a most probable value of 5·6 kcal g^{-1}.

In fact the mean calorific values of organic matter (5·6–5·8 kcal g^{-1} organic matter) agree with those given by American authors (Golley, 1961; Slobodkin and Richman, 1961; Comita and Schindler, 1963; Paine, 1964). It therefore appears to have been established with a fair degree of certainty that for most aquatic organisms the calorific value of organic matter approximates to 5·6–5·8 kcal g^{-1} organic matter. Deviations from this value are caused by variation

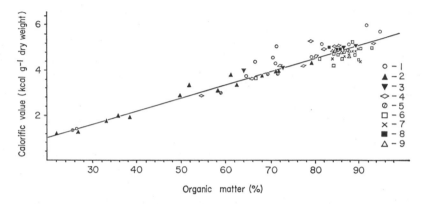

Fig. 2.2 The relationship of calorific value of aquatic organisms and their percentage organic content. 1—freshwater plankton according to Geng (Winberg *et al.*, 1934); 2—plankton from the Black and Azov seas (Vinogradova, 1956); 3—marine organisms (Vinogradova, 1956); 4—comparable crustacean data; 5—*Gammarus marinus*; 6—comparable data on Lamellibranchiata; 7—*Mytilus galloprovincialis*; 8—*Pachygrapsus marmoratus*; 9—*Pecten ponticus*. (4–9 from Ablyamitova–Vinogradova, 1949, 1949.)

in the chemical composition of the organic material and to some extent by changes in the ecological conditions and physiological state of the organisms (Slobodkin, 1961).

The relative proportions of proteins and carbohydrates has a relatively slight effect on calorific value because their energy equivalents are rather similar (5·65 and 4·10 kcal g^{-1}). A much greater influence on calorific value is produced by the lipid fraction in the bodies of aquatic organisms. The tissue fat content is closely related to an organism's ecological condition and physiological state, and sometimes may show great variation. Birge and Juday (1922) found that *Daphnia pulex* had a fat content of 21 per cent during the reproductive period but only 3·9 per cent at other times. Richman (1958) gives the calorific value of sexually immature Daphnia to be 5·05 kcal g^{-1} organic matter and that of mature individuals 6·09 kcal g^{-1}, that of the reproducing animals being 20 per cent higher.

Some aquatic organisms for various reasons exhibit considerable seasonal variation in their fat content. In some cases food is available for a restricted period of time and the animal must accumulate a sufficient quantity of energy-rich substance to provide it with energy reserves for the period when food is scarce (Conover, 1962). In other cases the seasonal variations in fat content and therefore in calorific value are associated with reproduction, development or other activities of the organism causing variation in fat content, and are studied seasonally (Fisher, 1962). Temperature also affects the fat content of a body to a considerable degree and this too results in seasonal fluctuations both in fat content and in calorific value.

Differences in chemical composition can significantly affect the calorific value of organic matter. If we consider two hypothetical cases, one with the maximum and the other with the minimum proportion of fat, the organic material in the former will consist of 80 per cent carbohydrate and 20 per cent protein, and in the latter 30 per cent protein and 70 per cent lipids. The calorific value of the former will be 4.4 kcal g^{-1} and that of the latter 8.3 kcal g^{-1}. Calorific values of 4.4 kcal g^{-1} or lower can be found in the literature. It is clear that such low values are due to erroneous determinations of either the calorific values or of the ash content. Any values below 5 kcal g^{-1} organic matter should be treated as suspect since very few organisms have such low values.

Variations in calorific value due to differences in chemical composition need to be taken into account by hydrobiologists, and the fact that many different kinds of organisms have an energy content of approximately 5.6 kcal g^{-1} organic matter, irrespective of species, makes it possible to use this value in very approximate calculations.

For more accurate data calorific values must be determined experimentally wherever possible.

2.3.2 *Direct Calorimetry*

Calorific values can be determined by direct calorimetry using an instrument called a bomb calorimeter. The basis of the method consists of combusting a sample in the presence of excess oxygen under high pressure inside the bomb which is immersed in water. The quantity of energy liberated due to the combustion is determined from the increase in water temperature. Complete combustion is ensured by compressing a weighed sample into a small pellet which is placed inside the bomb in contact with a fusible wire joining two electrodes. The bomb is filled with oxygen under a pressure of 25–30 atmospheres and placed inside the calorimetric vessel containing water. By applying a current to the electrodes, short-circuited by the fusible wire, the sample is ignited and combusted. Heat liberated during the burning causes an increase in the temperature of the water which is recorded by means of a special thermometer with a sensitivity of 0·002°C. From this measurement, and knowing the sample weight, the calorific value of the substance can be calculated.

Direct calorimetry is a method for determining the level of the so-called "physical" caloricity, that is the quantity of heat liberated during the complete oxidation of organic matter. If all the conditions of measurement are carefully observed the heat of combustion can be determined to an accuracy of 0·1 per cent (Drozdov, 1962). With biological material the accuracy of calorific value determinations is about 1 per cent (Faustov and Zotov, 1965). Because of its high repeatability, direct calorimetry is a method which is now frequently being adopted in hydrobiological and ecological studies.

The measurement of the ash content of a sample can be combined with the determination of calorific values by direct calorimetry, since the mineral fraction remains inside the bomb after the combustion of the sample. However, combustion of organic matter inside the bomb takes place at temperatures of 1000°C or so which are high enough to decompose certain salts. This could cause considerable error in any ash content determinations and must be taken into account. According to Paine (1964) this error may range from 5 to 33 per cent for ash contents of molluscan bodies. Unpublished data of Ostapenya demonstrates that ash content values of samples of marine plankton, determined from the residues inside the bomb, are undervalued by an average of 15 per cent. Clearly, erroneous determinations of the amount of mineral material present will distort any calculation of calorific value of the organic part of the sample, and endothermic processes accompanying the thermal decomposition of the salts can also affect the determined calorific values. From Paine's (1966) data the influence of this endothermic decomposition of salts is insignificant at ash contents of less than 25 per cent, but at levels of 50 per cent a correction of about 3–4 per cent ought to be applied to the determined calorific values.

It is not necessary to dwell in detail on the procedures involved in determining calorific values by direct calorimetry since every calorimeter has its own detailed instructions on how to carry out such determinations. When working with "Kreker", "Bertelo", SKB-52 or other similar bomb calorimeters commonly found in laboratories, best results are obtained with samples weighing between 0·5 and 1·5 g dry weight. The necessity for such large samples restricts the usefulness of direct calorimetry as a method for hydrobiological research. With microcalorimeters, samples can weigh as little as 50 mg (McEvan and Anderson, 1951; Slobodkin and Richman, 1960) which greatly increases the usefulness of direct calorimetry as a method. Unfortunately, microcalorimeters are not yet widely distributed in hydrobiological laboratories in the Soviet Union and are, therefore, in practice not available to most research workers.

2.3.3 *Wet Combustion Methods*
The basis of these methods consists of treating samples with a solution of a strong oxidizing agent such as $K_2Cr_2O_7$ or KIO_3. The difference between the initial quantity of oxidizing agent and the amount left after oxidation represents the amount of oxygen needed to oxidize the organic material in the sample. With this data, and with the help of oxycalorific coefficients, the calorific value of the sample can be determined.

The undoubted merit of wet combustion methods is the fact that, apart from their relative simplicity and general availability, calorific values can be determined with an accuracy of $\pm$ 3–5 per cent for samples of as little as 0·5–1·0 mg dry weight. This is considerably better than with present-day micro-bomb calorimeters.

Despite great differences in chemical composition, the quantity of energy per unit weight, or volume of oxygen, required for the complete oxidation of various organic substances (i.e. the oxycalorific coefficient) varies very little. Oxycalorific coefficients for protein, fat and carbohydrates are 3·305, 3·280 and 3·529 kcal g^{-1} oxygen respectively. Winberg et al. (1934) and Ivlev (1934) calculated the oxycalorific coefficients for various aquatic invertebrates and their results fell within very narrow limits, giving 3·33 to 3·49 kcal g^{-1} oxygen. A mean value of 3·38 kcal g^{-1} oxygen, or 3·4 kcal g^{-1} oxygen rounded off, is now generally adopted by hydrobiologists as the oxycalorific coefficient to use in wet combustion determinations of calorific value.

It is also possible to determine the weight of material oxidized by wet combustion methods. For this the oxygen equivalent per unit of organic substance must be known.

For calculating the quantity of material oxidized from its oxygen consumption, some investigators have used an oxygen equivalent of between 1·14 and 1·20 mg oxygen per mg organic substance. These values represent the oxygen equivalents of carbohydrates. Since the bodies of aquatic animals consist not only of carbohydrates, use of these equivalents must lead to erroneous results.

Skopintsev (1947) suggests adopting 1·50 mg oxygen per mg organic substance as an average oxygen equivalent for plankton, on the basis of published chemical compositions of marine plankton (Brandt and Raben, 1919) and freshwater plankton (Birge and Juday, 1922). This value has since been employed by various authors.

From what follows, this may be an underestimate. As has been mentioned, the calorific value of the organic material in aquatic organisms averages 5·6 kcal g^{-1} organic material and this value has been substantiated by several workers for a considerable amount of material. The oxycalorific coefficient of 3·38 kcal g^{-1} oxygen has not been disputed by anyone. By dividing the first value, which is the mean calorific value for plankton, by the second, it is possible to calculate an oxygen equivalent for organic matter and this comes to 1·65 mg oxygen per mg organic substance, which is 10 per cent higher than the oxygen equivalent proposed by Skopintsev.

Bichromate oxidation is the most popular of the wet combustion methods. In the Soviet Union this method was first introduced into limnological research by Winberg et al. (1934), and since then, with various modifications, it has become widely used both in fresh waters (Kuznetsov, 1945; Winberg and Zakharenkov, 1950; Winberg and Platova, 1951; Winberg, 1954; Votintsev, 1955; Nikolaeva and Skopintsev, 1961) and, lately, on the seas (Strickland, 1960; Sushchenya and Mikhalkovich, 1961; Sushchenya, 1961; Finenko, 1965; Ostapenya and Kovalevskaya, 1965).

Various modifications of chromate oxidation exist. Different authors recommend using a chromate solution with varying concentrations of bichromate. Thus Tyurin (1934) oxidized with a 0·4 N chromate solution and Ivlev (1934), a 0·2 N solution. Sivko (1960) tested the completeness of oxidation of organic material in river water with 0·1 N and 0·4 N chromate solutions and was unable to detect a difference. Maciolek (1962) obtained, to all intents and purposes, equivalent degrees of oxidation from chromate solutions of normalities between 2·00 and 0·05. It appears, therefore, that the concentration of the chromate solution may be any one within these limits. But it is necessary to remember that the error caused by inaccuracies in measurement of the oxidizing agent is increased in the more concentrated bichromate solutions.

The relative proportions of water and sulphuric acid in the oxidizing solution can have a great influence on the speed and completeness of oxidation. Friedemann and Kendall (1924) have shown that both the speed and completeness of the oxidative process is increased by using solutions with small amounts of water when oxidizing biological material. However, potassium bichromate in a chromic acid solution made up with concentrated sulphuric acid tends to decompose at temperatures above 150°C, and this has led to the development of two variant types of chromic acid solutions with potassium bichromate for oxidizing organic matter. One is a chromate solution made up with concentrated sulphuric acid and intended for use at a relatively low temperature (100°C) for 30 to 120 minutes. The other is prepared with dilute sulphuric acid and enables oxidation to be carried out at higher temperatures (180–200°C) for times of only 5 minutes.

The amount of oxygen consumed during oxidation of the sample is determined from the difference between the initial amount of $K_2Cr_2O_7$ and the amount left which does not react during oxidation. The quantity of bichromate remaining can be determined by various methods, the ones most usually employed being as follows.

1. Iodometric determination of potassium bichromate. When KI is added to the oxidizing mixture a quantity of iodine evolves which is equivalent to the bichromate present. The iodine is titrated against hyposulphite.

2. Potassium bichromate titrated against ferrous sulphate with diphenyl-amine as an indicator. The addition of a few millilitres of phosphoric acid to the chromate solution ensures a clear change from a reddish-blue colour to bright green by eliminating the influence of iron oxide ions.

3. Replacement of ferrous sulphate by the more stable solution of Mohr's salt in the titration. With diphenylamine as an indicator a sharp end-point is obtained with 0·2–0·1 N solutions of the ferrous sulphate or Mohr's salt (Nikolaeva, 1953). With such concentrated solutions, however, titration errors are considerable. Replacement of diphenylamine with phenylanthranilic acid (Simakov, 1957) ensures a good clear end-point in a 0·02 N solution of Mohr's salt; moreover, there is no need to use phosphoric acid to heighten the end-point colour change.

4. The quantity of bichromate in the solution can be determined photo-metrically by measuring the light absorption of the oxidizing solution at 440 mμ.

Thus the types of chromate oxidation method discussed differ in the conditions under which the oxidation reaction takes place—concentration of the oxidizing solution, temperature, and duration of heating—and also in how the results are recorded.

Soviet hydrobiologists have adopted two modifications of chromate oxidation. One of these is founded upon the method developed at the Kosino Limnological Station (Winberg et al., 1934). Oxidation is carried out in a 0·2 N solution of $K_2Cr_2O_7$ made up in concentrated sulphuric acid without a catalyst and heated in a water bath for two hours. The second modification is based upon the method of Tyurin (1934). Oxidation takes place in a 0·4 N solution of $K_2Cr_2O_7$ made up in sulphuric acid diluted 1:1 with water and boiled for 5 minutes in the presence of a catalyst.

Neither of these widely used modifications of chromate oxidation results in a complete oxidation of the organic matter found in the bodies of aquatic organisms, and the results obtained depend greatly on the chemical composition of the organic substance to be oxidized. The first variant of chromate oxidation mentioned above results in average underestimation of 25 per cent, due mainly to the incomplete oxidation of the proteins (Ivlev, 1934). The results from the second variant are also underestimated by about 28 per cent, mainly due to the incomplete oxidation of the lipids.

A much more complete oxidation of organic matter can be obtained by combustion in a potassium bichromate solution in concentrated sulphuric acid, together with considerable heating, although this is accompanied with marked and irregular decomposition of the potassium bichromate, as has already been mentioned. It appears to be possible to select a temperature of operation at which the organic substances are oxidized fairly completely but at which no decomposition of the potassium bichromate in the concentrated sulphuric acid takes place (Simakov and Tsiplenkov, 1961). From these results a modification of chromate oxidation has been developed with which calorific values of aquatic organisms can be determined with 90 per cent oxidation (Ostapenya, 1965).

A weighed sample of 0·5–4·0 mg of dry, carefully ground material is placed in a 100-ml combustion flask. Oxidation is carried out with 10 ml 0·1 N $K_2Cr_2O_7$ in concentrated sulphuric acid in the presence of a catalyst (100 mg Ag_2SO_4). The flasks are heated for 15 minutes at 140°C in a thermally regulated drying oven. The chamber of the oven is equipped with a simple mechanism for producing two opposing air currents in order to eliminate any temperature gradients present and to ensure an even heating of the samples. This mechanism consists of a fan set up on an axis inside the chamber and with a motor on the outside of the oven.

After being heated, the flasks are cooled and 15 ml of distilled water is added slowly, so that it washes down the necks of the flasks. This causes the contents to heat up considerably. After further cooling the excess bichromate is titrated against 0·02 N solution of Mohr's salt in the presence of phenyl-anthranilic acid.

This method of chromate oxidation retains all the initial advantages of Tyurin's procedure, as well as making the oxidation of organic matter more complete. It is the method recommended for determining calorific values of aquatic organisms.

Iodate oxidation can also be employed to determine calorific values of aquatic organisms (Karzinkin and Tarkovskaya, 1960; Birger, 1961; Makhmudov, 1964). Potassium iodate is a powerful oxidizing agent and, in an acidic medium in the presence of organic material, it will release oxygen which will oxidize the organic fraction of a sample.

Karzinkin and Tarkovskaya (1960) have proposed a modification of the procedure for iodate oxidation for determining the calorific content of small samples. A weighed sample of 8–15 mg of dry material is placed in a 300-ml flask. To the sample are added 3 ml of a 5 per cent solution of KIO_3 and 20 ml of concentrated sulphuric acid. The contents of the flask are boiled for 1 hour and, when cool, 50 ml distilled water are added to the now clear, yellowish fluid. Free iodine is evolved. The flask is warmed, but not boiled, until the colour and odour of iodine have completely disappeared, and its contents are diluted with 100 ml water. Then 10 ml of a 10 per cent solution of potassium iodide are added and the flask is placed in the dark for 10 minutes. The iodine now evolved is equivalent to the residue of potassium iodide not used during the oxidation; it is titrated with a 0·1 N solution of hyposulphite. A 90 per cent oxidation is obtained by this method, a level similar to that for the bichromate procedure just described. Both the iodate and bichromate oxidations are suitable methods for making calorimetric determinations of body tissues of aquatic organisms, although the bichromate method is more convenient when whole series of samples are being treated.

2.3.4 *The Calculation of the Calorific Value of a Sample from its Chemical Composition*

All organic substances have a characteristic combustion heat. This makes it relatively easy to calculate the calorific content of an organism once the chemical constituents of its organic fraction have been determined and the calorific equivalents of these are known.

It is usual to classify the organic substances forming the bodies of organisms into three major groups, fats, carbohydrates, and proteins. During combustion in a calorimeter these substances liberate the following average quantities of heat: fats 9·45, carbohydrates 4·10 and proteins 5·65 kcal g^{-1}. These mean values are cited in the well-known classic of Brody (1945) with references to the work of Fries, Atwater and Bryant, Benedict and Atwater, Rubner and a number of other authors. Different kinds of fats, carbohydrates and proteins have quite distinctive energy contents. Thus, for example, among the carbohydrates, glucose contains 3·8 kcal g^{-1}, sucrose 4·0 kcal g^{-1}, starch and glycogen 4·2 kcal g^{-1} (Kleiber, 1961). Some authors take 9·30 rather than 9·45 kcal g^{-1} as the calorific equivalent for fats, although the latter is to be found in the majority of publications (Brody, 1945; Richman, 1958; Kleiber, 1961).

If the constituents forming the organic material in a sample are expressed as percentages, then the calorific value of the sample can be calculated from the formula

$$\frac{5 \cdot 65P + 4 \cdot 10C + 9 \cdot 45F}{100} \text{ kcal } g^{-1} \qquad (2.1)$$

where P, C and F represent the percentage content of protein, carbohydrate and fat, respectively.

Where the calorific equivalents used in such calculations have been checked by combusting the proteins, fats and carbohydrates in a bomb calorimeter, the resulting calorific values must correspond to measured values obtained by direct calorimetry, that is, to the so-called physical caloricity.

During their combustion, both in the bomb calorimeter and in the living organism, fats and carbohydrates burn to the same products (CO_2 and H_2O) and therefore their physical calorific content is the same as their physiological one. Nevertheless, the physical and physiological calorific value of proteins and other nitrogenous compounds differs considerably because in living organisms the nitrogenous food substances are not oxidized completely and consequently the physiological calorific equivalent of nitrogenous compounds is lower than their physical one.

Animals belonging to different taxonomic groups excrete various final products of protein metabolism. Ammonia (NH_3) is the nitrogenous waste product for most aquatic invertebrates and freshwater fish. In many insects, in reptiles and in birds it is uric acid ($C_5H_4N_4O_5$) and amphibians, marine fish and mammals excrete mainly urea (N_2H_4CO) (Koshtoyants, 1950). These nitrogenous waste products possess different energy values and that is why the physiological calorific equivalent of protein compounds will vary for animals belonging to different groups.

Accepting the mean physical calorific value of protein to be 5·65 kcal g^{-1}, the physiologically available energy of protein will be 5·5 kcal g^{-1} in those animals which oxidize their protein to ammonia and in animals which excrete urea and uric acid it will be 4·8 and 4·3 kcal g^{-1} respectively.

Some authors consider that in mammals not only is urea excreted in the urine but also other substances such as creatine, and so the calorific equivalent of the protein they assimilate will be somewhat less than 4·8 kcal g^{-1}. Lusk (1928) and Brody (1945) consider it to be 4·3 kcal g^{-1}. Assuming that man assimilates protein from a mixed diet at an average efficiency of 85 per cent, each gramme of consumed protein liberates 4·1 kcal on oxidation (Rubner, 1902; König, 1904). This value has become firmly established in the literature on nutritional physiology, together with the values of 9·3 kcal g^{-1} for fats and 4·1 kcal g^{-1} for carbohydrates [Tables of chemical composition and nutritional value of food products, 1961 (Burshtein, 1963).]

These equivalents, often called the Rubner coefficients, have been accepted uncritically by many authors who have calculated the calorific content of aquatic organisms from their chemical composition (Kizewetter, 1954; Vinogradova, 1960, 1961; Malyarevskaya and Birger, 1965; Kostylev, 1965).

Undoubtedly, calorific values computed with the help of Rubner's coefficients will be lower than measured ones because the nitrogenous excretory product in aquatic invertebrates and freshwater fish is ammonia and not urea. The correction introduced by Rubner to cover the assimilation efficiency of protein in man cannot really be applied with any validity to aquatic animals, particularly as their assimilation efficiency appears to vary rather widely and, in any case, has not been studied thoroughly enough.

As has been demonstrated by Ostapenya (1968), the use of Rubner's coefficients to calculate the calorific content of aquatic organisms from their chemical composition results in underestimates of about 18 per cent compared with data obtained by direct calorimetry. On the other hand, calculations applying the coefficients 5·65 for proteins, 9·45 for fats and 4·10 for carbohydrates produce results which are virtually the same as those obtained by combusting samples in the bomb calorimeter.

Hence the use of coefficients which take account of the incomplete oxidation of proteins can only be justified where the calculated calorific value will serve as an index of the nutritional value of the food organism. The type of protein metabolism of the consumer ought also to be considered. Where the calculated calorific values are needed in order to express biomass or individual weight in units of energy, those coefficients should be used which correspond to the energy released when proteins, fats and carbohydrates are combusted in a bomb calorimeter.

2.3.5 *A Comparison of Methods for Determining Calorific Values*

A comparison of all the methods described above for determining calorific value, reveals that the most accurate results are those obtained by direct calorimetry. The possibilities of this method are greatly reduced by the fact that the apparatus required is not generally available to the laboratories that need it. Also the procedure is relatively complex, which makes it a cumbersome method when many samples have to be burnt.

Wet combustion is a comparatively simple method, convenient for serial determinations and applicable for samples of small weight but, because of the incompleteness of oxidation, the calorific values obtained represent only 90 per cent of those derived from direct calorimetric measurements. An adequate correction has to be introduced in order to obtain realistic values.

An advantage of wet combustion methods is the possibility of determining approximately the weight of substance oxidized over and above the main aim of obtaining the substance's calorific content.

The method of calculating calorific content from the chemical composition gives only approximate values. The results obtained are influenced by the errors associated with determining the chemical composition and are dependent on whether the conventional energy equivalents for proteins, fats and carbohydrates are applied.

2.4 Analysis of Proteins, Fats and Carbohydrates in Organic Matter

2.4.1 *Determination of Proteins*

In hydrobiological research the protein contained in aquatic organisms is usually determined from the nitrogen content as established by the method of

Kjeldahl. The amount of protein is obtained by multiplying the measured amount of nitrogen by 6·25. The nitrogen content in various proteins varies between 13 and 19 per cent and the coefficient $6·25 = 100/16$ is a conventional value related to an average nitrogen content of 16 per cent. Kjeldahl's method determines not only proteinous nitrogen but also the nitrogen contained in other organic compounds, and so calculation of the amount of protein from the nitrogen content gives only approximate values, usually called "crude protein". The possible error is ± 20 per cent.

Photometric methods are now being quite extensively used for determining the chemical composition of naturally occurring organic compounds and they are beginning to be used in hydrobiology. The greatest virtue of photometric methods is the possibility they offer for working, at a satisfactory level of accuracy, with samples of only a few milligrammes dry weight.

Among photometric methods available for determining the protein contained in the bodies of aquatic animals, some attention should be drawn to the simple and straightforward biuret reaction which is based on the reaction of proteins with $CuSO_4$ in an alkaline medium (Krey, 1958; Racusen and Johnston, 1961; Anan'ichev, 1961).

Streickland and Parsons (1960) recommend Keller's method (Keller, 1959) which is based upon a modified Elson-Morgan reaction which is described in detail in their manual. It is to be noted that it is about six times more sensitive than the biuret method.

More recently Lowry's method has been widely used for determining proteins. This method is based on a combination of the biuret reaction and the phenol reagent of Folin (Lowry et al., 1951). The procedure is as follows. The sample of organic substance is subjected to alkali while being warmed in a water bath; to the resulting solution, first an alkaline solution of copper sulphate and then Folin's reagent are added. The solution becomes blue in colour, the intensity of which is proportional to the protein concentration of the sample. This is one of the most sensitive and accurate methods of determining protein at present available. Both the ninhydrin reaction and direct spectrophotometric measurement at 280 mμ are less accurate than Lowry's method and the biuret reaction is 100 times less sensitive.

Lowry's method for protein determination requires the following reagents: (1) A—a 2 per cent solution of Na_2CO_3 in 0·1 N NaOH; (2) B—a 0·5 per cent solution of $CuSO_4.5H_2O$ in a 1 per cent solution of sodium tartrate; (3) C— a mixture of 1 ml of reagent B and 50 ml of reagent A; (4) Folin's reagent[1];

---

[1] To prepare Folin's reagent, into a 1·5-litre flask put 100 g of $Na_2WO_4.2H_2O$ and 25 g $NaMoO_4.2H_2O$ and add 50 ml of 85 per cent H_3PO_4 and 100 ml concentrated HCl.

(5) 1 N solution of NaOH in a flask connected to a condenser and boiled for 10 hours. After boiling, 150 g of lithium sulphate plus 50 ml water are added together with a few drops of liquid bromine. This mixture is boiled for 15 minutes, the flask being disconnected from the condenser, in order to eliminate excess bromine. The cooled contents are then adjusted to 1 litre with distilled water and filtered. The final reagent should show no green coloration.

To determine protein in dried material by Lowry's method, the procedure is as follows (Ostapenya, 1964): into a volumetric centrifuge tube put 1–2 mg of material finely ground in a mortar, and add 5 ml 1 N NaOH. The tubes are then heated in a boiling water bath for 10 minutes. According to Hewitt (1958), prolonged heating does not improve the protein extraction. After heating the tubes are cooled and their contents adjusted to a volume of 10 ml with distilled water. The samples are then centrifuged for 5 minutes at 3000 rpm. Into a small tube (20–30 ml) are placed 5 ml of reagent C and 1 ml of the supernatant. These are vigorously stirred and left for 10 minutes, after which, to each tube, 0·5 ml Folin's reagent, diluted with an equal volume of water, is added. After 40 minutes, the samples are analysed photometrically at 750 mμ. The protein content of the sample is obtained by applying the measured optical density to a calibration curve.

In this method of determining protein the relationship between optical density and protein concentration is non-linear (Lowry *et al.*, 1951). In preparing the calibration curve great care should be taken to use only purified casein. According to Lowry, for protein determined by this method from whole unextracted tissue errors can be about 3 5 per cent.

Determination of protein by the methods just examined can be recommended for bulk determination of protein from the bodies of aquatic organisms. The choice of method has to be left to the judgement of the worker himself. It is useful to remember that the biuret reaction is simple and readily performed. For samples involving micro-quantities of material Lowry's method is preferable.

2.4.2 *Determination of Fats*
The most frequently employed method for determining the fat content of aquatic organisms is that of fat extraction in a Soxlet apparatus followed by weighing of the extracted lipids. Soxlet's method is well enough known for any description of its procedure to be unnecessary.

The various methods of fat extraction depend upon good infusion. A weighed sample of the substance being examined is soaked in a solvent and left to infuse for some time (several periods of 24 hours). A known sub-sample of the solvent, together with the dissolved lipids, is transferred to a tared vessel. As the solvent evaporates the vessel is dried to a constant weight and, from the increase in weight of the tared vessel, the fat content of the tissue can be calculated. There are various modifications of the infusion technique which differ for the solvent used (sulphuric ether, chloroform, petroleum ether, mixtures of various solvents) and in the duration of soaking. Extraction is mostly used for macro-samples. Vityuk (1964) has developed a procedure for determining lipids from micro-samples by infusion. In this method a sample (3–10 mg) of dried, finely ground material is weighed on a micro-analytical balance sensitive to 10^{-5} grammes, placed into a pycnometer and soaked in dry sulphuric ether. The samples are then left to extract for a

week, during which time the contents of the pycnometer are stirred carefully twice a day. After seven days of infusion the volume of the extract is adjusted to 5 ml and a 2 ml sub-sample is sucked up in a pipette from the pycnometer and transferred into a tared vessel for evaporation. The dishes are dried at 40°C under thermally regulated conditions until they reach a constant weight and, from the increased weight of the vessels, the quantity of extracted fat can be calculated. Sample weights between 2·34 and 19·09 mg do not influence the results obtained. According to Vityuk, parallel samples deviate from the mean by between 0·15 and 3·36 per cent. On the whole this method is quite success-ful. However, the necessity of weighing samples to an accuracy of 10^{-5} mg reduces its possible use in many hydrobiological laboratories possessing only ordinary balances.

FIG. 2.3
Apparatus for
fat extraction.

Strickland and Parsons (1960) determined fat by a method proposed by Mukerjee (1956). The lipids are washed in an alka-line solution of ethyl alcohol and then treated with a pinacyanol reagent. The coloured complex which develops is extracted with bromobenzol. At 620 mμ, the extinction of the blue-coloured extract is measured, and a cali-bration curve is prepared with standard solutions of stearic acid. The main disadvantage of this method is the light-sensitivity of the coloured complex. Moreover, the pH greatly influences the fat levels determined after the addition of pinacyanol; the optimal pH is $8·7 \pm 0·1$.

A rapid method for determining the fat content of small samples is the technique described by Ostapenya (1964) in which the fat fraction of the sample is extracted in a simple apparatus (Fig. 2.3). The main portion consists of a conical centrifuge tube (10 ml) with an indentation around its "waist". Resting upon this "waist" is a conical extraction tube made of filter paper, which is absolutely free of both fat and ash. The carefully ground and weighed sample is placed inside the paper extraction tube and this is placed in the glass tube previously weighed to a constant weight and containing 2–3 ml of petroleum ether. A finger-shaped condenser, readily constructed from another test tube, is inserted into the top of the centrifuge tube.

The apparatus is heated on an air bath at 70°C. The vapours of the petroleum ether condense on the condenser, providing a uninterrupted supply of fresh drops of solvent to extract the lipids from the sample material. When extraction is completed, both the condenser and the extraction tube are removed, the solvent is allowed to evaporate and the centrifuge tube, together with the lipid fraction, is weighed to a constant weight. The amount of lipids extracted from the weighed sample is determined from the gain in weight of the centrifuge tube.

Samples of about 10–15 mg dry weight must be used in order to use this method for lipid determination. All weighing must be carried out on an analytical balance with a sensitivity of 10^{-4} g. Duration of extraction is 6 hours. For sample weights of 15 mg deviations from the mean of parallel determinations did not exceed $\pm$ 7 per cent. This method can be recommended for extensive series of lipid determinations in small samples.

2.4.3 *Determination of Carbohydrates*

When investigating the chemical composition of aquatic organisms, many authors obtain the carbohydrate content indirectly by difference from 100 per cent of the summed percentage contents of protein, fat and ash, all determined experimentally. The values so obtained reflect the true carbohydrate content only very approximately since they also incorporate the sum of all the errors inherent in the methods for determination of proteins, fats and ash. However, as no very satisfactory procedure exists for determining carbohydrates with an adequate degree of accuracy in small samples, for a long time carbohydrate content could only be determined by percentage difference. Recently, as a result of the development of photometric techniques for biochemical analyses, simple and accurate procedures have appeared which can be used successfully for the determination of the carbohydrate content of aquatic organisms.

The method for determining carbohydrates with the anthrone reagent in concentrated sulphuric acid has become very popular in biochemical research. Hewitt (1958) used this method to determine carbohydrates in *Chlorella* cells. Strickland and Parsons (1960), in their manual, also recommend this method for carbohydrate determinations of aquatic animals.

A weighed sample (1–3 mg) of finely ground material is suspended in 10 ml distilled water. To 1 ml of this suspension 10 ml of anthrone reagent (0·2 g anthrone, 8 ml absolute ethyl alcohol, 30 ml water and 100 ml concentrated sulphuric acid) is added. The contents of the test tubes are well mixed and heated for 7 minutes in a boiling-water bath. The tubes are immediately cooled and analysed photometrically at 620 mμ. The calibration curve is prepared with glucose.

Mendel *et al.* (1954) have proposed a simple micro-method for determining the concentration of blood carbohydrates which has also proved suitable for carbohydrate determination in aquatic organisms (Raymont and Krishnaswamy, 1960; Raymont and Conover, 1961; Ostapenya, 1964).

The basis of this method lies in the fact that dilute solutions of glucose or other similar carbohydrates, when in contact with concentrated sulphuric acid, develop 5-hydro-oxymethylfurfurol which interacts with the substances present to give a bluish-rosy colouring. The reaction is specific for glucose, fructose, glycogen, starch and other carbohydrates.

Carbohydrate determinations by this method require only two reagents: a 5 per cent solution of trichloroacetic acid, containing 0·1 per cent Ag_2SO_4 and chemically pure concentrated sulphuric acid with a specific gravity of 1·84. Carbohydrates are extracted from the weighed sample (2–6 mg dry weight) in the trichloroacetic acid and heated for 30 minutes in a water bath. During this procedure, proteins and chlorides, which interfere with the determination, are also precipitated during the extraction. After completion of the extraction the samples are centrifuged at 3000 rpm for 5 minutes and 3 ml concentrated sulphuric acid is added to 1 ml of the clear supernatant. After vigorous shaking the mixture is warmed for precisely 6·5 minutes in a boiling-water bath. The samples are then cooled and analysed photometrically at 520 mμ. Glucose is used to prepare a calibration curve. The possible error is ± 5 per cent.

This method is simple, convenient and can be recommended for determining the carbohydrate content of aquatic organisms.

Chapter 3

GROWTH, RATE OF DEVELOPMENT AND FECUNDITY IN RELATION TO ENVIRONMENTAL CONDITIONS

3.1 General Principles of Animal Growth

The production of the population of a species consists of the sum of the growth increments of the individuals forming that population, including the growth due to their sexual products and other organic substances which become separated from the body during the period being considered. That is why any estimation of production requires quantitative data about growth, duration of development of different stages, and fecundity, as well as how these properties are influenced by environmental conditions. From this follows a need for general concepts about types of growth and the relation between duration of development and fecundity and temperature and other environmental factors.

Individual growth is a process of increase in mass in a developing organism. The study of growth processes involves following changes either in body weight (or some other measure proportional to weight, such as nitrogen content) to give growth in weight or in linear dimensions to give growth in length. It is clear that patterns of growth in weight are fundamentally important because they are related to types of metabolism.

In practice, what is most frequently measured is growth in length, so that its relationship to growth in weight must first be clarified.

The relationship between any index of a linear dimension of a growing animal l and its weight w may be expressed by the equation

$$w = q l^b \qquad (3.1)$$

where q is a constant, equal to w when $l = 1$.

Where the geometric proportions of the body are not changed, that is, there is no alteration in the body form, then $b = 3$. If the body form changes during growth so that the ratio of the linear measurement to weight decreases, then $b > 3$ or, in the opposite case, $b < 3$ (Table 3.1).

When values of the constant q are being determined from empirical measurements, equation 3.1 must be used in its logarithmic form

$$\lg w = \lg q + b \lg l.$$

C

33

As can be seen, lg w and lg l are linearly related to each other (Fig. 3.1). Empirical data fit this formula when they are distributed along a straight line on a logarithmic graph. In such a plot, lg q is the intercept on the y-axis where lg $l = 0$ and b is the regression coefficient of lg l on lg w and may be easily calculated by one of the appropriate statistical methods. Whenever equation 3.1 is used, the range of lengths l and weights w from which the values of q and b have been obtained and for which they are valid must be specified.

TABLE 3.1

Examples of length–weight coefficients in freshwater hydrobionts

Organism	q	b	Range of weights (mg)	Authors
Daphnia spp.	0·052	3·0	0·01–5·0	Pechen', 1965
Bosmina spp.	0·124	2·2	0·002–0·2	Pechen', 1965
Diaphanosoma	0·092	2·4	0·02–0·2	Pechen', 1965
Macrocyclops albidus	0·055	2·7	0·001–0·25	Klekowski and Shushkina, 1966
Asplanchna priodonta	0·209	3·0	0·0008–0·25	Bregman, 1968
Moina sp.	0·081	3·0	0·01–0·2	Kryuchkova and Kondratyuk, 1966
Einfeldia carbonaria	0·018	2·2	0·3–3·8	Gavrilov and Arabina, 1967
Sphaerium corneum	0·669	2·6	5·0–259·0	Gavrilov and Arabina, 1967
Asellus aquaticus	0·133	2·4	0·9–81·5	Gavrilov and Arabina, 1967
Chironomus plumosus	0·034	2·1	1·1–85·0	Gavrilov and Arabina, 1967
Anodonta anatina	0·052	3·0	5·0–30·0	Gavrilov and Arabina, 1967

Note. The values for b presented in the table were obtained under certain conditions, and cannot be considered as characteristic for every population of the species but depend, within certain limits, upon experimental conditions, type of fixative, etc.

Equation 3.1 is applicable to a wide range of representatives in the animal kingdom (fish, molluscs, crustaceans, insect larvae) and often the values of b obtained do not differ greatly from 3, but are usually between 2·5 and 3·5. Some authors (Konstantinov, 1962) believe that $b = 3$ can be used for a wide range of animals, but this tolerates rough approximations in calculation.

There is no good reason for expressing the relationship between linear dimensions and weight by means of other empirical formulae, which in any case provide no advantages. Thus, Kamshilov (1951) expresses the relationship between cephalothorax length of marine copepods to their weight by the equation $w(\text{mg}) = (0·286l(\text{mm})+0·05)^3$. This relationship can be expressed by the practically equivalent equation

$$w = 0·0242 l^{2·984}.$$

The following basic concepts are involved in any examination of the quantitative laws of the process of growth.

Absolute growth increment per unit time Δt—Δw: Δt. When Δt is small enough the absolute increment becomes the growth rate dw/dt.

Specific growth increment, that is, growth per unit weight, is given by

$$(1/w)(\Delta w/\Delta t).$$

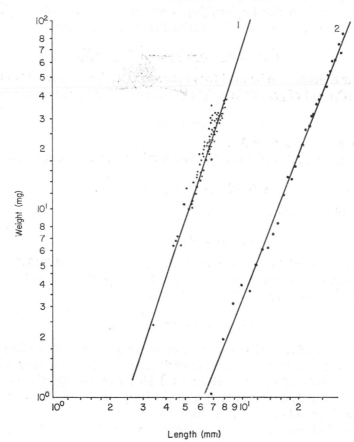

FIG. 3.1 Examples of length–weight regressions in animals plotted on double logarithmic scales.

1—*Anodonta anatina*; 2—*Chironomus, plumosus*.

When Δw is made small enough we obtain the specific rate of growth, usually represented by C_w:

$$C_w = (1/w)(\Delta w/\Delta t).$$

For all types of growth, the mean specific rate of growth $\bar{C}_w$ for a period $t_1 - t_2$ can be computed from the equation

$$\bar{C}_w = (\ln w_2 - \ln w_1)/(t_2 - t_1) = (\lg w_2 - \lg w_1)/[0 \cdot 4343(t_2 - t_1)] \quad (3.2)$$

where w_2 and w_1 are the weights at the end and beginning of the period, that is, the moments of time t_2 and t_1.

When the mean specific growth rate is not large and the period $t_2 - t_1$ is relatively small, so that $(w_2 - w_1)/w_1$ is not greater than 0·10, a sufficiently accurate estimation of it can be obtained from the equation

$$\bar{C}_w = 2(w_2 - w_1)/[(w_2 + w_1)(t_2 - t_1)].$$

Similar concepts can be used for linear growth. The relative rates of linear growth (C_l) and of growth in weight (C_w) are related by the simple ratio

$$C_l = C_w/b$$

where b is the index from 3.1.

When growth takes place at a constant relative rate, that is, when

$$(1/w)(dw/dt) = C_w = \text{constant},$$

weight and time are exponentially related and

$$w_t = w_0 e^{C_w t} = w_0 10^{0 \cdot 4343 C_w t} \quad (3.3)$$

where w_t is the weight at time t, w_0 is the initial weight and $0 \cdot 4343 = \log e$.

With this type of growth a straight line is obtained on a semi-logarithmic graph, since $\lg w$ and t are linearly related so that

$$\lg w_t = \lg w_0 + 0 \cdot 4343 C_w t.$$

Some types of animal growth (insect larval growth, embryonic growth, first stages of post-embryonic growth of fish) may be described approximately by this formula of exponential growth. However, usually the individual growth of animals decreases in rate as the body size increases and this may be better described by equations considered later.

In biological productivity studies the properties of individual growth must be considered together with intensity of metabolism.

In the simple case mentioned above, where growth takes place at a constant rate, the rate of metabolism is independent of the weight of the animal. In fact, among animals of the most varied kind, the metabolic rate and relative growth rate both decrease with increase in body size.

It is well known that the relationship between energy metabolism and individual weight of vertebrates, invertebrates and even one-celled animals may be expressed by a power function, which in biological literature is often known as the parabolic or allometric equation, i.e.

$$Q = Mw^{a/b}, \quad Q/w = Mw^{-(1-a/b)} \tag{3.4}$$

where Q is the total metabolism (for example, the quantity of oxygen consumed by an individual per unit time), Q/w is the rate of metabolism (metabolism per unit weight), M is a constant equal to the total metabolism of an animal of unit metabolic size $w^{a/b}$, w is individual weight and a/b is a constant such that $0 < a/b < 1$. The value for a/b usually lies within the limits 0·6–0·9.

Equation 3.4 can be stated as

$$T = T_1 w^{a/b} \tag{3.5}$$

where T is the metabolic loss of one individual per unit time, expressed in the same units as w, and T_1 is a constant equal to the metabolic loss of an individual of unit metabolic size.

The conversion of 3.4 to 3.5 can be illustrated by an example. It is known that the oxycalorific coefficient is approximately equal to 3·4 cal mg^{-1} or 4·86 cal ml^{-1} oxygen. From this the rate of energy dissipation in metabolism can be obtained from the rate of oxygen consumption. For example, $M = 0\cdot3$ ml oxygen g^{-1} h^{-1} is equivalent to $0\cdot3 \times 4\cdot86 = 1\cdot458$ cal g^{-1} h^{-1}. To calculate metabolic loss in units of weight it is necessary to know the calories per unit body weight. If the calorific value is 600 cal g^{-1} fresh weight, and $w = 1$ g, then, in our example, the metabolic loss T_1 comes to

$$1\cdot458/600 = 0\cdot00243 \text{ g}$$

and, for twenty-four hours, 0·0583 g or 5·83 per cent of w, equalling 1 g.

When the ratio v between the growth increment P and metabolic loss T remains constant during growth, then $P = vT$. This ratio $P/T = v$ shows a simple relationship to the coefficient of utilization of assimilated food for growth. From $K_2 = P/(P+T)$ it follows that $v = K_2/(1-K_2)$.

Substituting that $P = dw/dt$, and bearing in mind equation 3.5, we find that

$$dw/dt = vT_1 w^{a/b} = Nw^{a/b} \tag{3.6}$$

where $N = T_1 v$.

The expression 3.6 is the equation of "parabolic growth" in its differential form. It therefore follows from the above that growth must be parabolic in situations where v and K_2 remain constant. However, the opposite is not necessarily true since growth may be parabolic not only when K_2 remains constant but also if K_2 is a power function of w.[1] Thus, knowing growth, metabolism can be predicted provided K_2 is known.

---

[1] This point was made by Zaika in June 1969 at the Symposium held at the Lake Naroch Biological Station.

Parabolic growth formulae have been used to describe the growth of many organisms (insect larvae, certain crustaceans and, in particular instances, fish and other animals).

Equation 3.6 shows the relationship of growth increment to weight already achieved. The integral form of this equation shows the weight as a function of time, i.e.

$$w = [N(1-a/b)(t-t_0)+w_0^{1-a/b}]^{b/(b-a)} \tag{3.7}$$

where w_0 is the initial weight at time t_0 and w is the weight attained during the period $t-t_0$.

Taking that $1-a/b = n$, we obtain

$$w^n - w_0^n = Nn(t-t_0). \tag{3.7a}$$

It is useful to note that

$$N = (w_2^n - w_1^n)/[n(t_2-t_1)]$$

where w_1 and w_2 are weights corresponding to the periods of time t_1-t_0 and t_2-t_0.

If time is measured from an initial point when $w = 0$, then from 3.7a we obtain

$$w = (Nnt)^{1/n}. \tag{3.8}$$

In order to apply this equation to growth data expressed in the form given in 3.7, it is necessary to determine t_0. In the case where $w = 0$ and $t = 0$, from 3.7a we find that

$$t_0 = w_0^n/Nn.$$

In parabolic growth, as w increases, the relative growth rate C_w decreases proportionally to w^{-n}:

$$C_w = (1/w)(dw/dt) = Nw^{-(1-a/b)} = Nw^{-n}. \tag{3.9}$$

From 3.8 we see that, if time is measured from the moment when $w = 0$, then

$$N = w^n/nt.$$

By comparison with 3.9, we have for this case

$$C_w = 1/nt = t^{-1}b/(b-a), \tag{3.9a}$$

that is, the specific growth rate, when it is of a parabolic type, is inversely proportional to time t, provided that time is measured from the moment when $w = 0$.

In spite of this decline in its specific rate with time, parabolic growth is represented by a concave curve, since a/b is always less than 1. For higher values of the exponent a/b (when a/b is closer to unity), and consequently when n is small and $1/n$ is large, the formula for parabolic growth may be usefully applied to situations where, for all practical purposes, growth may be regarded as exponential, especially for portions of the growth curve corresponding to sufficiently large values of t.

The rate of parabolic growth decreases with time because of the decrease in rate of metabolism with increasing size, but w increases without limit. However, in real examples of parabolic growth, as in insects, growth suddenly ceases when the imaginal stage is reached.

The formula for parabolic growth was obtained from equation 3.4 and the coefficient K_2 via a generalization of the physiological data relating metabolism and weight. This opens up the possibility of carrying out the reverse computation. Having obtained the value for the constant N from observations on growth, and knowing K_2, it is possible to calculate T_1 from the expression

$$N = T_1 v = T_1 K_2 / (1 - K_2).$$

Having calculated T_1, we can obtain the value for metabolic loss T from 3.5.

For many animals the growth rate decreases more rapidly than in the case of the parabolic type of growth and tends to zero as the definitive size and weight is approached. In this situation the growth curve has an S-shaped form. With this kind of growth the rate decreases not only because of the decline in metabolic rate with increase in weight but also because the relationship of growth rate to metabolic loss does not remain constant but declines. It is clear, therefore, that in growth of this type, K_2 and v attain their maximum values, $(K_2)_m$ and $v_m = (K_2)_m / [1 - (K_2)_m]$, at the beginning of growth. Furthermore, v decreases proportionally to some function of the weight already attained: $v = v_m f(w)$. Consequently from 3.6, the general growth formula corresponding to the definitive sizes when $N = T_1 v_m$ may be represented as

$$\mathrm{d}w/\mathrm{d}t = T_1 v_m w^{a/b} f(w) = N w^{a/b} f(w). \tag{3.10}$$

In special cases, where $f(w) = (W^n - w^n)W^{-n}$, that is when v decreases in proportion to $W^n - w^n$,

$$\mathrm{d}w/\mathrm{d}t = T_1 v_m [(W^n - w^n)/W^n] w^{1-n}. \tag{3.11}$$

Taking note that $1 - n = a/b$, because $n = 1 - a/b$, thus

$$v = P/T = K_2/(1 - K_2) = \{(K_2)_m / [1 - (K_2)_m]\}\{(W^n - w^n)/W^n\}. \tag{3.11a}$$

This last expression has a special significance since, by integrating 3.11 and taking into consideration that $N/W^n = k$, we can obtain the well-known growth equation developed by Pütter, Bertalanffy and Taylor (Winberg, 1966):

$$w = [W^n - (W^n - w_0^n)e^{-nk(t-t_0)}]^{1/n}. \tag{3.12}$$

It has been shown by many authors that this equation can usefully represent growth data for a variety of animals.

In applying equation 3.12, it is useful to know that, if W and n are known, k may be determined from, for example, two empirically obtained points on

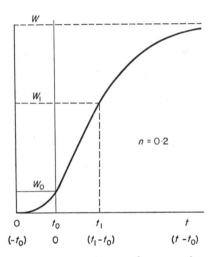

FIG. 3.2 A growth curve from equations 3.12 and 3.12a.

the growth curve, one of which may be w_0. If w_1 and w_2, as well as their corresponding periods $t_1 - t_0$ and $t_2 - t_0$, are known, then

$$k = \frac{1}{n(t_2 - t_1)} \ln\left(\frac{W^n - w_2^n}{W^n - w_1^n}\right) = \frac{2 \cdot 3026}{n(t_2 - t_1)} \lg\left(\frac{W^n - w_2^n}{W^n - w_1^n}\right).$$

Equation 3.12 can be stated in a more convenient and simpler form if time is measured from the moment when $w = 0$, that is when $t = 0$. In that case

$$w = W(1 - e^{-nkt})^{1/n}. \tag{3.12a}$$

Before changing over from 3.12 to 3.12a it is necessary to determine the value for t_0 (Fig. 3.2). Then, provided that $t = 0$ when $w = 0$, from 3.12 we find that

$$t_0 = \frac{1}{nk} \ln\left(\frac{W^n}{W^n - w_0^n}\right) = \frac{2 \cdot 3026}{nk} \lg\left(\frac{W^n}{W^n - w_0^n}\right).$$

If w is expressed as a fraction of W and the product kt is taken as our independent variable, then growth expressed as w/W is dependent only on n:

$$w/W = (1-e^{-nkt})^{1/n}. \tag{3.12b}$$

It is an easy matter to plot a series of growth curves from equation 3.12b corresponding to selected values of n (Fig. 3.3). From 3.12b it is also possible

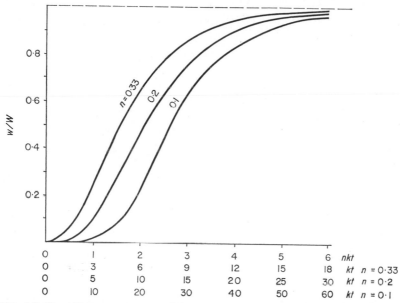

FIG. 3.3 Growth curves constructed from equation 3.12b for different values of n.

to determine at what value of kt weight w reaches some predetermined fraction of W. Moreover, knowing kt, and if t can be obtained from empirical data, then the corresponding value for k can be found, or, if the value for k is obtainable, the value for t can be found.

The value for k, and thus too for N, may also be determined by other methods; for example from measured values of the relative growth rate C_w corresponding to selected values of w. Bearing in mind that $w^{a/b}w^{-1} = w^{-n}$, we can see from 3.11 that for the type of growth being considered C_w depends on w, i.e.

$$C_w = (1/w)(dw/dt) = Nw^{-n}(W^n-w^n)/W^n = N(w^{-n}-W^{-n}) = k[(w^n/W)^{-n}-1]. \tag{3.13}$$

It can be clearly seen that the relationship C_w/N for each given w/W depends only on n, so that

$$C_w/N = w^{-n}-W^{-n} = W^{-n}[(w/W)^{-n}-1] \tag{3.13a}$$

For known values of n it is possible to find the corresponding values of the ratio C_w/N for any magnitude of w/W and w.

If for a given value of w, such as the mean weight of a certain size class, we can empirically determine the specific growth rate C_w', then the magnitude of C_w/N corresponding to this value of w can be obtained from the expression 3.13a and it then becomes possible to calculate N from

$$C_w'(C_w/N)^{-1} = N. \tag{3.13b}$$

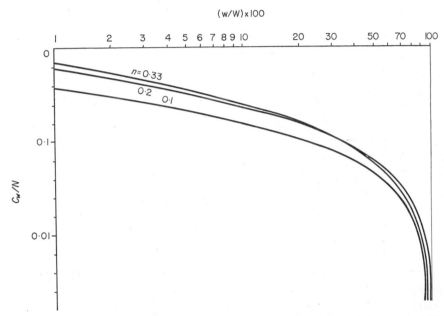

FIG. 3.4 The relation of the ratio between specific growth rate C_w and the constant N to the equation 3.12a for various values of n.

Figure 3.4 shows that the ratio C_w/N falls off rapidly in the region of the definitive weight and as C_w approaches zero. From the same Figure we see that the ratio C_w/N is only weakly dependent on n.

In many situations it is useful to know how the ratio of growth increment to metabolic loss P/T and the coefficient of utilization of assimilated food for growth K_2 decline with increase in weight. From 3.11a we see that

$$P/T = K_2/(1 - K_2) = v_m[1 - (w/W)^n]. \tag{3.14}$$

From which it follows that

$$K_2 = \{v_m[1 - (w/W)^n]^{-1} + 1\}^{-1}. \tag{3.14a}$$

Recalling that

$$v_m = (K_2)_m/[1-(K_2)_m]$$

and substituting from 3.12a, we obtain

$$1-w^n/W^n = e^{-nkt}.$$

We may represent P/T and K_2 as functions of time, thus

$$P/T = v_m e^{-nkt} \tag{3.15}$$

and

$$K_2 = (v_m^{-1}e^{nkt}+1)^{-1}. \tag{3.15a}$$

All the above growth indices have been introduced with some serious reservations. Growth has been considered as increase in body weight only whereas, during the reproductive period of many animals, the development of the sexual products among the females constitutes a large part of the total growth increment or may become the sole growth phenomenon after a certain stage. Among many animals, such as the females of freshwater planktonic copepods, growth practically ceases when sexual maturity is attained when only increase in body weight is considered, whereas it continues at a rapid rate when the matter and energy of the liberated eggs are also taken into account. Pechen' and Kuznetsova (1966) found that the total dry weight of eggs laid by one female *Daphnia pulex* exceeded 2·5 times her final weight (Fig. 3.5), thus demonstrating how large the relative growth rate in the form of eggs can be.

It is natural to assume that the growth increment of the eggs per unit time contributes a certain fixed fraction v_g of the metabolic loss, that is $P_g = Tv_g$, so that $P_g = T_1 v_g w^{a/b}$. Taking the simplest situation, where v_g is constant, since one knows how w varies with time, it is not difficult to find how P_g changes as growth proceeds.

It is more realistic to postulate that the relation between egg growth and metabolic loss decreases. Then v_g does not remain constant but decreases with time. At present, because of lack of data, it is still not possible to derive a mathematical expression which would describe this decrease, and special studies are needed to determine the value of v_g. For each given weight w for which the corresponding value of P_g is known, v_g can be calculated from the formula

$$v_g = P_g/(T_1 w^{a/b}) = P_g/T. \tag{3.16}$$

The total growth consists of the individual increase in size, $P = dw/dt$, plus P_g, so that, when the food requirements are assessed from the coefficient of utilization of assimilated food energy for growth, realistic values of K_2 must be computed by taking into account loss due to egg growth, as in the formula

$$(P+P_g)/T = v+v_g = K_2/(1-K_2) \tag{3.16a}$$

r

$$K_2 = (v+v_g)/(1+v+v_g).$$

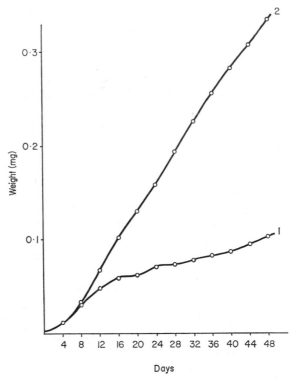

FIG. 3.5 The individual growth of a female *Daphnia pulex*. 1—body weight without sexual products; 2—body weight together with the cumulated weight of all the eggs hatched during a given period.

The question frequently arises as to how to incorporate into the growth equations discussed above the different conditions, and especially food conditions, which are known to affect growth rate.

Trophic and other factors can affect the growth rate, mainly via the metabolic rate which is influenced by external conditions. To what extent and in what way conditions for growth can affect the value of K_2 is not yet very well known. It is clear, however, that with inadequate food conditions a major part of the food assimilated goes to cover metabolism, so that K_2 becomes smaller. Consequently, the constant N, which for a given w is proportional to the relative growth rate, must also have a lower value under unfavourable conditions compared with optimal conditions.

In addition, the definitive weight W is also in some degree related to trophic and other growth conditions.

In practice, we often have to deal with data on the growth of individuals belonging to age or size groups that fall short of the definitive size, and this, together with the ever-present scatter of empirically observed points, complicates any assessment of which of the three possible influences on the parameters of the growth equations is most important for a particular species. Probably further study will demonstrate that the situation differs for various species under different conditions.

It is also necessary to remember the nature of the assumptions made, namely that in growth which follows the S-shaped curve, both its rate and the ratio P/T decrease in proportion to $W^n - w^n$. This assumption can be justified by the fact that the earlier well-known growth equation (3.12), which has been shown many times to represent satisfactorily the growth of many species, can be derived from the S-shaped growth equation. This very important circumstance should not be undervalued; nor should we forget that we are considering only one of the special forms of the general equation 3.10. There are no biological reasons for rejecting its other forms.

From 3.10, we find that

$$C_w = T_1 w^{-n} f(w).$$

This expression makes it possible to carry out special investigations on growth under more constant conditions in order to show how $f(w)$ should be expressed for different species. It will be necessary to determine C_w empirically for various values of w and to establish for which forms of $f(w)$ is the condition that N is constant satisfied. As mentioned earlier, equations 3.8 to 3.12 apply only in those cases where the stated conditions of $N = $ constant satisfies

$$f(w) = v_m(W^n - w^n)W^{-n}.$$

In applying the concepts developed above to actual data, we meet with many difficulties and complications which arise from large differences in growth of different individuals in a population, from the abrupt changes that occur during the course of development of particular stages (e.g. in metamorphosis), from seasonal variation in growth or from the variable external conditions and many other causes.

Nevertheless, none of these circumstances, which make the actual course of growth of individual species much more complex than the simple picture considered above, eliminates the possibility of discovering the basic forms and general laws of growth. On the contrary, the complex course of growth observed in any particular animal can be interpreted and understood only if the general laws of growth are known and the special conditions affecting the growth of a given species are made clear.

In applying these growth equations, it is very necessary to keep clearly in mind within what limits and under what conditions they may be used. It is evident that closest agreement with the ideal course of growth following a known law can be obtained for species with relatively simple patterns of growth when the whole cycle of growth and development takes place under constant conditions. This can only mean experimental work in the laboratory. What are urgently needed are simultaneous studies on the growth, metabolism and feeding of each species being investigated. An extensive development of such work would provide a firm basis for understanding the more complex phenomena of growth.

The selection of a particular procedure for working up data on growth cannot be made merely on the basis that some parts of the growth curve obtained approximate to a known equation. Such a purely formal basis is quite inadequate for a fruitful application of the quantitative laws of growth. Let us illustrate this important assertion by means of an example. Suppose that it is known only that a certain portion of the left-hand concave part of a growth curve agrees strictly with the expression 3.12a. Because of the usual scatter of empirical points, this portion can also be approximated by 3.7, which implies that we accept the postulate that $w_0 \neq 0$ and has some positive value. Then the relative growth rate for the weight w will be equal to

$$C_w = Nw^{-n}$$

whereas, under the conditions associated with equation 3.12a, C_w must be expressed as

$$C_w = Nw^{-n} - NW^{-n}.$$

If, for the determination of N, we take the empirically obtained value for C_w', then, depending on which of the two equations are used for the calculation, we will obtain different values of N from 3.13b and consequently different metabolic losses, compared with those more approximate values obtained from 3.7.

By examining in this way the relationships between growth and metabolism, growth equations can be carefully fitted with the empirical data which reflects the actual biological features of the species being studied.

3.2 The Dependence of Developmental Rate on Temperature

All estimates of the production of aquatic animals involve knowing the duration of the developmental period (embryonic period, larval life, etc.). The more exactly this is known, the more adequately and precisely may production be estimated. As is known, the duration of poikilotherm development depends upon temperature, whose range in nature may be very wide. Thus, various appropriate "temperature corrections" need to be applied to our production estimates. For this purpose it is necessary to have some idea about the nature of the relationship between temperature and the duration of development.

The results of many biological studies have shown that the relationship between temperature and duration of development has the form illustrated as curve (a) in Fig. 3.6. The prolongation of development at the highest temperatures was observed only under laboratory conditions and probably is not significant ecologically, since at such near-lethal upper temperatures in nature neither normal development nor survival for any length of time is likely. Consequently, any mathematical expression relating development and temperature should not incorporate this part of the curve (as did Janisch, 1932).

In studies on the influence of various factors on development it is advisable to make use of, not the duration of development D, but its inverse, $v = 1/D$, which is the developmental rate. This rate has the dimensions of time to the power minus one and tells us what fraction of the total development takes place per unit time.

Ignoring the descending part of the curve on the upper right (the dashed line in Fig. 3.6, curve (b)), this graph of developmental rate as a function of temperature looks like an S-shaped curve, with lower left concave and upper right convex sections. The mathematical expression for a curve of this shape is rather complex. But the situation is eased by the fact that most estimates of production are based upon measurements of developmental rate in the middle zone of temperature, which is not too different from the optimal zone. However, the boundaries of the temperature zone within which normal development occurs are not immediately obvious and can only be determined as the upper and lower temperatures at which 50 per cent of the individuals survive. Such experimental determination of temperature limits are best carried out by means of probity tests, described in statistical textbooks.

In those situations where the optimal temperature zone lies in the middle section of the S-curve, the ascending straight section of curve (b) in Fig. 3.6 (the solid line) describes approximately the relation of v to temperature. The well-known "rule of sum of temperatures"[1] can be applied conveniently

---

[1] This name, "the rule of sum of temperatures", is not quite exact: it should be "the rule of sum of degree-days or degree-hours".

to this middle section in order to estimate the "temperature correction". Many authors (Bodenheimer, 1934; Peairs, 1927; Blunck, 1923) have applied this rule in its quantitative form.

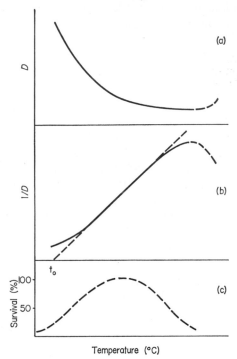

FIG. 3.6 Some general types of temperature relations.
(a)—duration of development D; (b)—rate of development $1/D$; (c)—percentage survival.

In accordance with the rule of sum of temperatures, the product of the duration of development D and the effective temperature $t-t_0$ gives the value of the constant:

$$S = D(t-t_0) \qquad (3.17)$$

where t_0 is the temperature of the conventional "biological zero" or the "lower thermal limit", corresponding to the point where the straight line describing developmental rate at various temperatures intersects the abscissa. From 3.17 we see that

$$v = S^{-1}(t-t_0) = (St)^{-1}-(St_0)^{-1}. \qquad (3.18)$$

When $v = 0$, then $t = t_0$.

In order to determine t_0 and S, it is enough to know the two values D_1 and D_2 for the corresponding temperatures t_1 and t_2. Then

$$t_0 = (D_1 t_1 - D_2 t_2)/(D_1 - D_2).$$

When several values of $1/D_1$, $1/D_2$, $1/D_3$ ... corresponding to the temperatures t_1, t_2, t_3 ... are known and when graphically all the points lie on a straight line, then it is possible to calculate the regression coefficient (3.18) which best fits the empirical results, using one of the known methods (method of averages, method of least squares, graphically, etc.). From 3.17, based on the rule of sum of temperatures, it can be seen that the duration of development is related hyperbolically to temperature,

$$D = S(t - t_0)^{-1}. \tag{3.19}$$

The constant S represents the product of time and effective temperature (degree-days, degree-hours, etc.).

The value of S is characteristic for each species. The larger it is, the longer is the duration of development for a given effective temperature. Its formal meaning can be stated as the developmental duration at 1°C effective temperature which has no direct biological significance, since lower effective temperatures result in unrealistic values for v, calculated from the rule of sum of temperatures.

The inverse value, $1/S = k$, called the "coefficient of thermal lability" by certain authors (Mednikov, 1965; Kozhanchikov, 1946), corresponds to the regression coefficient of rate of development $1/D$ on temperature. The formal meaning of k was considered to be the fraction of development per unit time at 1°C effective temperature.

The rule of sum of temperatures has been applied many times to the embryonic and larval development of insects, using wide intervals of mean temperatures. For instance, Konstantinov (1958) determined the duration of larval development of certain species of chironomids for the temperatures 15, 20 and 25°C, publishing the following results for the total number of degree-days and the t_0 characteristic for each species:

	S	t_0 (°C)
Chironomus dorsalis	306–350	6
Chironomus annularis	390–395	7
Chironomus plumosus	615–640	5
Polypedilum nubeculosum	238–252	8
Cryptochironomus pararostratum	208–229	9
Limnochironomus nervosus	330–382	9

This rule may be useful for providing an approximate expression of the relation of developmental rate to temperature in other animals. That this is so is suggested by Fig. 3.7, in which are plotted curves relating the embryonic developmental rate of some freshwater crustaceans to temperature. These are based upon the experimental results of Elster and Eichhorn, from which

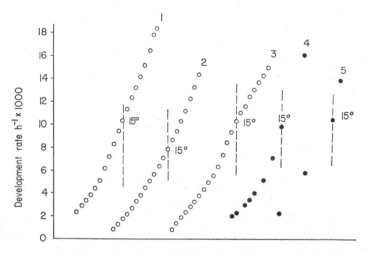

FIG. 3.7 The rate of embryonic development of freshwater crustaceans in relation to temperature. Abscissal divisions consist of equal intervals of 5°C.

1—*Eudiaptomus gracilis* (Elster, 1954); 2—*Acanthodiaptomus denticornis* (Eichhorn, 1957); 3—*Myxodiaptomus lacinatus* (Eichhorn, 1957); 4—*Daphnia galeata mendotae* (Hall, 1964); 5—*Daphnia longispina sevanica eulimentica* (Meshkova, 1952). 1, 2 and 3—points taken from curves constructed from empirical data; 4 and 5—empirical data.

they constructed their tables of developmental durations, which were obtained without much material, making it meaningless to attempt to interpret the differences in the form of the curves 1, 2 and 3.[1] It is interesting that in the curve for *Diaptomus gracilis* the straight section falls between those temperatures which, according to Elster, are similar to those of the water the species inhabits during the summer.

The concave or convex sections of the curve relating rate of development to temperature, which cannot be approximated to a straight line, may be expressed by the formula of Bělěradeck (1929, 1935) which differs from 3.19

---

[1] Both Elster and Eichhorn, from their own data, came to erroneous conclusions about the inconstancy of the sum of degree-hours. This explains the incorrect calculations of these values. Attention was not paid to the necessity of determining t_0 and the temperatures used were the measured experimental ones and not the effective temperature $(t-t_0)$.

in that a power index b is applied to $t-t_0$ ($b>1$ for a concave curve; $b<1$ for a convex curve).

$$D = a/(t-t_0)^b \quad \text{or} \quad 1/D = a^{-1}(t-t_0)^b \qquad (3.20)$$

According to this formula, the relation between a rate process and temperature is expressed with the aid of not one but two constants. This greatly complicates the application of this quite empirical formula, rendering it unsuitable for comparisons so that it is not recommended, although recently it has had its advocates (McLaren, 1963).

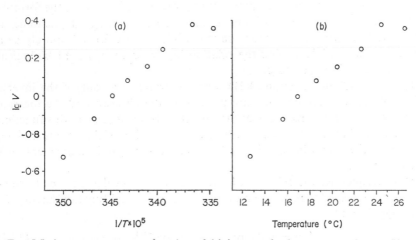

FIG. 3.8 A rate process as a function of (a) inverse absolute temperature, or (b) temperature on the division rate in *Paramecium caudatum*. On the ordinate, log number of divisions per 24 hours (Mitchell, 1929).

Two exponential functions, with only one constant, are available for relating the rate of metabolism and other biological phenomena to temperature. These functions are more or less different but they express the acceleration due to temperature of the growth process. They are useful for the left concave section of the curve relating the rate of development to temperature. The rate process is considered as an exponential function of either inverse absolute temperature $1/T$ or temperature t (Fig. 3.8).

The first of these two functions is the well-known formula of van't Hoff–Arrhenius:

$$v = v_0 e^{-(\Delta F/R)(1/T)} \qquad (3.21)$$

where $\Delta F/T$ is the value with the same dimensions (cal mol^{-1} deg^{-1}) as R which is the gas constant with a value 1·986 cal mol^{-1} deg^{-1}. Therefore, the ratio $\Delta F/RT$ is a coefficient without dimension and indicates the acceleration of a process with increase in temperature. ΔF, when applied to biological processes, is called the "temperature characteristic", often designated μ or E. In order to relate the rates of a process at two temperatures t_2 and t_1, we may use

$$v_1/v_2 = e^{-(\Delta F/R)(1/T_2 - 1/T_1)}. \tag{3.21a}$$

The original physicochemical meaning of ΔF is that it was the activation energy of the chemical reaction whose rate was being studied. If it is considered that the influence of temperature upon metabolic and developmental rates reflects the relation of temperature to protein synthesis, then the formula of van't Hoff–Arrhenius has some significance for biological phenomena. Recently, however, most authors have considered this formula purely as a convenient way of describing the relationship between observed biological phenomena and temperature.

The graph of this function (3.21) forms a straight line only if the inverse values of absolute temperature are plotted on the abscissa and the log rate of the process on the ordinate (see (a), Fig. 3.8). For this reason, this function can be applied only in those cases where the disposition of the empirical data, plotted on the graph in such a system of coordinates, is in agreement with this condition.

More frequently, the relation of the rate of biological processes to temperature is revealed with the help of another exponential function having the following form:

$$v = v_0 \alpha^t \quad \text{or} \quad v_2/v_1 = \alpha^{(t_2 - t_1)}. \tag{3.22}$$

This function forms a straight line in a semi-logarithmic graph, as

$$\lg v_2 = \lg v_1 + (t_2 - t_1) \lg \alpha.$$

Consequently, it expresses more adequately those empirical results which show a linear relationship when plotted with temperature on the abscissa and log rate of the process on the ordinate (see (b), Fig. 3.8).

The function 3.22 is employed in two forms.

1. Assuming that $\alpha = e^k$, we obtain the formula of Tauti (1925):

$$v_2/v_1 = D_1/D_2 = e^{k(t_2 - t_1)}. \tag{3.23}$$

The constant

$$k = v^{-1} dv/dt$$

(Mednikov's coefficient of thermal lability) is a more precise expression of the relation of the rate process to temperature,

$$k = (\lg v_2 - \lg v_1)/[0{\cdot}4343(t_2 - t_1)]. \tag{3.24}$$

2. Assuming that $\alpha = Q_{10}^{1/10}$, then

$$v_2/v_1 = D_1/D_2 = Q_{10}^{(t_2-t_1)/10}. \tag{3.25}$$

In this case we have the well-known and widely used formula usually called the "van't Hoff temperature coefficient". The numerical value of this coefficient directly reveals the magnitude of the acceleration of a process per 10°C increase in temperature.

It is clear that

$$\lg Q_{10} = 10(\lg v_2 - \lg v_1)/(t_2 - t_1) \tag{3.26}$$

and Q_{10} and k are related so that

$$\ln Q_{10} = 2 \cdot 3026 \lg Q_{10} = 10k. \tag{3.27}$$

When certain values of v_1, v_2, v_3 corresponding to the temperatures t_1, t_2, t_3 are known, then, since the functions 3.23 and 3.25 form a straight line on a semi-logarithmic graph, the values for k and Q_{10} may be determined by one of the generally accepted methods for estimating the parameters of a straight line.

Often there arises the necessity to compare data expressed as ΔF, as Q_{10} or k.

Noting that $t_2 - t_1 = T_2 - T_1$ in 3.21a and 3.23, we can see that

$$k = \Delta F/(1 \cdot 986 T_1 T_2) \tag{3.28}$$

and, from 3.21a and 3.25, that

$$\lg Q_{10} = 2 \cdot 187 \Delta F/T_1 T_2. \tag{3.29}$$

Table 3.2 facilitates the comparison of ΔF, Q_{10} and k and has been compiled for two temperature intervals.

As we can see from the formulae 3.28 and 3.29 and the Table, the relation between Q_{10}, k and F depends on which temperature interval is involved; for example, if $\Delta F = 15\ 500$ for the temperature interval 5–15°C, the corresponding value of $Q_{10} = 2 \cdot 63$, whereas $Q_{10} = 2 \cdot 46$ for the temperature interval 15–25°C.

The real significance of these differences is not great since in only a few cases, with very wide temperature intervals, are ΔF, Q_{10} or k constant. As a rule, rates of biological processes are more dependent on temperature at lower temperatures than at higher ones. This fact is reflected in the empirical "normal curve" of Krogh (1914, 1916) which was constructed from respiratory measurements in different species of animals and has been used very widely ever since. The dependence of a rate process on temperature, according

TABLE 3.2

A table comparing the values of F, Q_{10} and k at two temperatures

$F = \mu$ (cal mol^{-1})	$t = 5-15°C$		$t = 15-25°C$	
	k	Q_{10}	k	Q_{10}
6 000	0·0374	1·45	0·0349	1·42
8 000	0·0499	1·65	0·0466	1·59
10 000	0·0624	1·87	0·0582	1·79
12 000	0·0749	2·11	0·0698	2·01
14 000	0·0874	2·39	0·0815	2·26
15 500*	0·0967	2·63	0·0902	2·46
17 000	0·1061	2·88	0·0989	2·69
18 500	0·1154	3·15	0·1077	2·91
20 000	0·1248	3·49	0·1164	3·19

* Values of F, k and Q_{10} approximating to Krogh's "normal curve".

to Krogh's curve, can be calculated for separate temperature intervals by employing the following coefficients:

	5–10°C	10–15°C	15–20°C	20–25°C	25–30°C
Q_{10}	3·5	2·9	2·5	2·3	2·2
k	0·125	0·106	0·091	0·083	0·079
ΔF	19 570	17 230	15 350	14 440	14 130

For values of the correction on the basis of this curve, see Table 3.3 which provides relative values of the ordinate q for various temperatures and may serve for conversions to rate units at 20°C.

TABLE 3.3

Values of the temperature correction factor (q) for converting respiratory rates to 20°C, according to the "normal curve" (Winberg, 1956)

t	q	t	q	t	q	t	q
5	5·19	12	2·16	19	1·09	26	0·609
6	4·55	13	1·94	20	1·00	27	0·563
7	3·98	14	1·74	21	0·920	28	0·520
8	3·48	15	1·57	22	0·847	29	0·481
9	3·05	16	1·43	23	0·779	30	0·444
10	2·67	17	1·31	24	0·717		
11	2·40	18	1·20	25	0·659		

Where the rate of development follows the rule of sum of temperatures, then it is easily seen from 3.18 that

$$v_2/v_1 = (t_2 - t_1)/(t_1 - t_0). \tag{3.30}$$

Consequently, in these cases, each interval of effective temperature has a corresponding well-defined acceleration of developmental rate, i.e. well-defined values of ΔF, Q_{10} or k. This suggests that the degree of acceleration of developmental rate is not related to the inherited specific development rate which is given by the value of the constant S in degree-days or degree-hours.

According to 3.4 and 3.25, we can obtain the following values for separate intervals of effective temperature:

$$\text{from } (t_1 - t_0) \text{ to } (t_2 - t_1) \quad 5\text{–}10°C \quad 10\text{–}15°C \quad 15\text{–}20°C$$
$$Q_{10} \qquad\qquad\quad 4·0 \qquad 2·25 \qquad 1·78$$

As can be seen, in those cases where the rule of sum of temperatures is applicable, the acceleration of developmental rate with increased temperature declines rapidly from low to high temperatures, whereas it increases very quickly at effective temperatures lower than 5°C. It is necessary to remember that in this region the rule of sum of temperatures no longer reflects the real relation of development to temperature. The actual curve of this relationship is concave in form at low temperatures and somewhat left of the straight line (see curve (b), Fig. 3.6). Assuming that t_0 falls most frequently in the temperature interval 5–10°C, and using the rule of sum for the middle temperatures, we obtain an acceleration of development similar to Krogh's curve.

This is a fact of great significance since the rule of sum of temperatures is a generalization based upon empirical data relating the developmental rates of a great number of animals to temperature.

We must ask to what extent empirical data corresponds to the above ideas. It is difficult to say anything about this, as the relevant information is widely dispersed throughout biological literature and has yet to be gathered together and generalized. The bulk of this information is concerned with embryonic development and temperature and very few long-term continuous observations are available on developmental rates of the feeding stages of organisms, which are, of course, the stages most interesting to us. The speed of development of feeding animals, particularly at higher temperatures, is a function not only of metabolic rate, as in non-feeding animals, but also of food consumption. It is known that food consumption begins to fall off somewhere between the optimal temperature and the upper limit of tolerated temperatures, that is, nowhere near the temperature of most intense metabolism. Thus, maximal rates in filter-feeding, even in warm-water species, occur at relatively low temperatures (Fig. 3.9).

Consequently it is to be expected that acceleration of both growth and developmental rates of feeding animals will decline rapidly at high tempera-

tures. Although special studies are needed to clarify this question, the data
which are available do not conflict with these expectations. Some results of
Brown (1929) are relevant here; he outlined the relation of cladoceran developmental rates to temperature and measured the duration of the period from

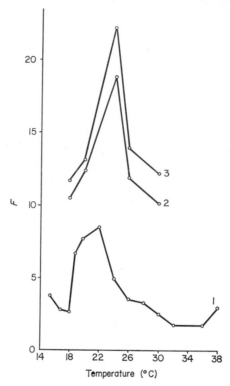

Fig. 3.9 Filtration rate (F ml per individual in 24 hours) in relation to temperature
(°C).
1—*Moina rectirostris*; 2—*Simocephalus
vetulus*; 3—*Daphnia pulex* (Kryuchkova-
Kondratyuk 1966).

birth to the first appearance of young, from which he calculated the following
values of Q_{10}:

	13–20°C	20–30°C	30–35°C
Moina macrocopa	3·96	2·39	1·76
Pseudosida bidentata	2·90	2·03	1·63
Simocephalus spp.	2·36	1·81	1·15
Daphnia pulex	2·64	1·70	—

It is interesting that this data agrees with the rule of sum of temperatures (Fig. 3.10).

A very important result was obtained by Mednikov (1962), who used already published data on developmental durations of marine, brackish and

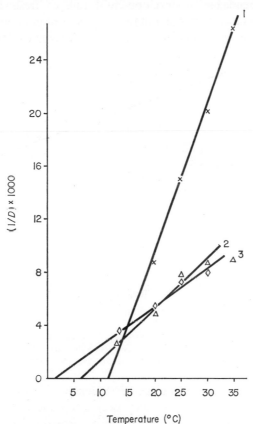

Temperature (°C)

FIG. 3.10 Developmental rate of cladocerans in relation to temperature.
D—duration of development from egg to sexual maturity in hours. 1—*Moina macrocopa*; 2—*Simocephalus* spp.; 3—*Daphnia longispina* (Brown, 1929).

freshwater planktonic copepods at their optimal temperatures to produce one general equation relating developmental rate and temperature which is applicable to all the species (Fig. 3.11):

$$D = 125e^{-0.0833t}$$

where D is the duration of development in days at temperature t. It should be noted that $k = 0\cdot0833$, which corresponds to a value of Q_{10} equal to $2\cdot30$, again reveals a level of acceleration of development due to temperature which generally agrees with the expression given above for the middle temperature region.

A similar dependence of developmental rate of freshwater planktonic crustaceans on temperature was obtained by Shushkina (1964) and she has commented on the agreement of her results with Krogh's "normal curve".

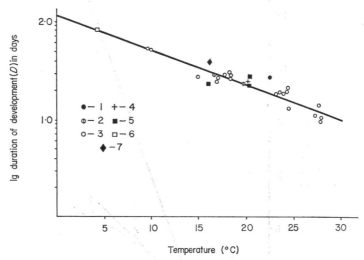

FIG. 3.11 The duration of metamorphosis of calanoids at various temperatures.
1—*Acartia clausi*; 2—*Centropages kroyeri*; 3—*Calanipeda aquaedulcis*; 4—*Pseudodiaptomus coronatus*; 5—*Neutrodiaptomus incongruens*; 6—*Calanus finmarchicus*; 7—*Acanthodiaptomus denticornis* (Mednikov, 1962).

Consequently, even in the absence of sufficient data to characterize how the developmental rate of a particular species is influenced by temperature, it is possible either to apply a "temperature correction" derived from the "normal curve" or to make use of the fact that, for the temperature interval 15–25°C, $Q_{10} = 2\cdot25$, $k = 0\cdot081$ and $\Delta F = 13\,820$.

Generally this practice will not result in too great an error, especially when the temperature interval is small and its mid-point is used. But, when the upper or lower temperature limits are approached, or with species which normally inhabit low or high temperature conditions, some care must be exercised.

It is generally acknowledged that there exists both a physiological and an inherited capacity to adapt to the temperature conditions in which organisms live.

To some degree in many aquatic animals, adaptation or acclimation to low temperature intensifies metabolism, and therefore the developmental rate, whereas the reverse is true in high-temperature adaptation, both processes being depressed. This phenomenon may be revealed by comparing the developmental rates of several populations of a species, each living at a different mean temperature. Such a comparison may disclose some populations whose development is only slightly affected by temperature. Mednikov suggests that populations of the branchiopod *Triops cancriformis* are a good example of this phenomenon.

It is possible or even probable that temperature adaptation has a seasonal effect on how temperature influences developmental rate. For example, it is not an accident that the developmental rate of *Diaptomus graciloides*, measured at Lake Erken at 7°C by Nauwerck (1963), was significantly higher than was expected.

In considering our order of priority in this field of investigation it is clear that we need (*a*) to critically appraise and generalize data already published on development and temperature and (*b*) to clarify the relationship between temperature and the developmental rate of the embryonic and post-embryonic stages of the dominant species of aquatic animals by comparing experimental results on growth and developmental rates with those determined under field conditions.

3.3 The Effect of Food and Temperature on the Fecundity and Reproductive Rates of Poikilotherms

To estimate the population production of animal species, we need to know the reproductive rate and fecundity of animals living under various trophic and temperature conditions.

How these factors influence developmental rate and fecundity is not well known but, nevertheless, it is useful to examine the few results that do exist.

The intensity of animal reproduction, determined by both the mean egg number per brood and how frequently these are laid, may be expressed as the mean number of eggs laid by one female per unit time. Fecundity is defined as the mean number of eggs per brood.[1]

Temperature and food influence the fecundity of animals under natural conditions, mainly via their effects on body size since fecundity is directly related to the body size of the egg-producing females. This phenomenon

---

[1] Fixed material is often used in studies on fecundity but such material (usually fixed in 4 per cent formalin) may result in too low values for fecundity where, as in Cladocera, embryos fall out of the brood pouches. It is suggested that cladoceran fecundity should be determined from living specimens, or Hall (1964) recommends using a better fixative such as 95 per cent spirit.

has been demonstrated by Marshall (1949) for marine copepods (Fig. 3.12), by Ravera and Tonolli (1956) for two freshwater diaptomids, by Margalef (1955) for *Cyclops*, and by Galkovskaya and Lyakhnovich (1966), Dunke (1960) and Green (1954) for cladocerans.

The relation between body size and brood size (i.e. number of eggs) is expressible as the power function, $F - ml^k$, where F is the number of eggs per brood, l is the body length and m and k are constants. Consequently, those factors that affect animal body size and growth rates are the factors that also influence fecundity. Judging from published data, it is temperature and body size in planktonic animals which have the most marked effect upon fecundity.

Several authors (Coker, 1933; McLaren, 1965) have shown for the marine copepod, *Pseudocalanus*, for the chaetognath, *Sagitta*, and for freshwater

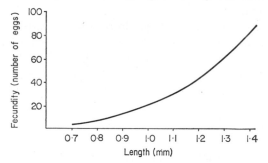

Fig. 3.12 Relation between brood size and body length in female *Pseudocalanus* (McLaren, 1965).

cyclopoids, that size under natural conditions is an inverse function of the mean habitat temperature (Fig. 3.13). The form of the McLaren (1965) curve relating egg numbers and body length of *Pseudocalanus* with habitat temperature demonstrates the characteristic relationship of fecundity and temperature (see Fig. 3.14). In these animals food quantity has some influence on the number of eggs but this effect is so slight (Deevey, 1960) that it can be ignored; thus, temperature appears to be the main factor affecting fecundity.

On the other hand experiments carried out by Hall (1964) demonstrated a direct relationship between the body size of adult *Daphnia galeata mendotae* and food concentration (Fig. 3.15). As the body size changed, so did the animal's fecundity. In this case the decisive factor was food. The possibility cannot be ruled out that the responses of these two groups of animals to temperature or food factors are diverse. McLaren (1965), considering the main indices relating the development of poikilotherms, temperature and food conditions, writes, ". . . growth rates and final sizes are functions of

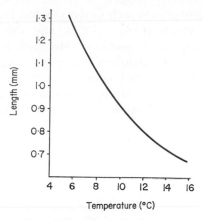

Fig. 3.13 Size of adult female *Pseudocalanus* as a function of temperature (McLaren, 1965).

food supply, while their development rates may be relatively unaffected. This is true of some species that have been most commonly used in experimental studies of growth and production (for example, *Daphnia* spp. and brine shrimp). However, there is another kind of poikilotherm in which size at any developmental stage (including maturity and final stage) is a function of temperature during development, although growth and development may be thwarted by food shortage. This sort of growth, far from being unusual, may

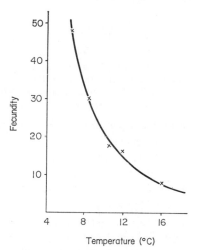

Fig. 3.14 Relation between fecundity and temperature in *Pseudocalanus* (as constructed from Figs. 3.12 and 3.13).

be the rule in important groups such as fish (Taylor, 1958), shellfish (Taylor, 1960), cyclopoid copepods (Coker, 1933), and calanoid copepods and chaetognaths among the marine plankton (McLaren, 1963)."

During seasonal (May to October) changes in fecundity of planktonic crustaceans, the spring maximum coincides with the period when water temperature is relatively low and the concentration of phytoplankton is high. It is quite possible that maximal fecundity is caused by different factors

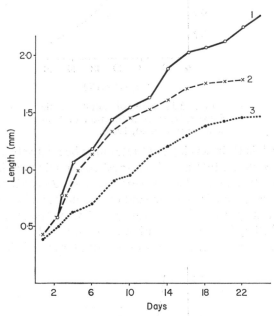

FIG. 3.15 Growth of *Daphnia galeata mendotae* in different concentrations of food (Hall, 1964).
1—1 mln. kl/ml; 2—0·062 mln. kl/ml; 3—0·015 mln. kl/ml; of protococcoid mixture.

in the various groups of animals present, i.e. low temperature in cyclopoids and food availability in cladocerans.

Increased fecundity with increase in food concentration has been demonstrated in experimental animals. Such results have been recorded for various groups of aquatic animals: Cladocera, mainly *Daphnia* (Ingle, Wood and Banta, 1937; Zhukova, 1953; Richman, 1958); Copepoda, *Diaptomus* (Nauwerck, 1963); Oligochaeta, *Enchytraeus* (Ivleva, 1953).

On the other hand, under natural conditions where two or more factors may operate simultaneously, cases are known in which fecundity has been closely correlated with food concentration (Meshkova, 1952; Comita and Anderson, 1959; Nauwerck, 1963).

It seems reasonable that animal fecundity is related to food concentration up to a critical level. Some general ideas about the nature of the relationship can be obtained from Fig. 3.16, which shows Green's (1954) results for *Daphnia magna*. Curves of this form can be described approximately by the function $f = F(1 - e^{-kC})$, where f is the fecundity at food concentration C, F is maximal fecundity and k is a constant. The food concentration associated with maximum fecundity will differ in ecologically distinct species. Hence the maximum fecundity of *D. cucullata*, a species normally living in low food concentrations, is likely to occur at a lower level of food than that for

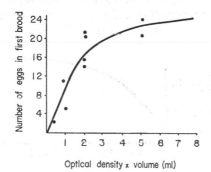

FIG. 3.16 Relation between fecundity of *Daphnia magna* and food quantity (*Chlorella*—Green (1954)).

D. magna which usually lives in abundant food conditions. One of the most important reasons for studying fecundity is to determine the food levels at which maximum fecundity occurs in species differing ecologically.

The reproductive rate B^1, derived from fecundity and expressed as the mean number of eggs laid by one female per day, can be calculated from the formula

$$B^1 = E/D$$

where E is the mean number of eggs laid by one female and D is the duration in days of embryonic development. Recently, values of B^1, computed in this way, have been employed to estimate the production of planktonic animals (Elster, 1954, 1955; Edmondson, Comita and Anderson, 1962; Hall 1964; Edmondson, 1960, 1962, 1965; Wright, 1965)—see Section 5.8.

As the value of B^1 is derived from fecundity, numbers of animals and the duration of their embryonic development, B^1 must reflect the dependence of these three parameters on temperature and nutritive conditions. Whilst fecundity and its relative value E tend to decrease as temperature rises to levels higher than optimal, simultaneously, the duration of embryonic development D shortens. In such a relationship with temperature, the value of $B^1 = E/D$ found is determined by E or D, whichever is more influenced by temperature. If, as is usual, D is more dependent on temperature than E, then B^1 will increase as temperature rises.

Many authors (Ingle, Wood and Banta, 1937; Zhukova, 1953; Elster, 1955; Shushkina, 1964b) have revealed that the duration of embryonic development is independent of nutritive conditions of females belonging to different species. Population numbers are involved in the calculation of E, but the reproductive rate depends not on the numbers of females but on the age structure of the whole population. Therefore the reproductive rate B^1 can be correctly used for production estimates only when the age structure of the population is constant. When such a situation exists, the relationship

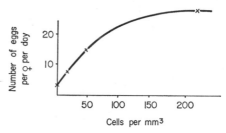

FIG. 3.17 Mean number of eggs laid daily by female *Calanus* in different concentrations of *Chlamydomonas* sp. (Marshall and Orr, 1952).

between the reproductive rate and nutritive conditions reflects the character of the relation between fecundity and feeding conditions.

In fact, in their experiments on *Calanus finmarchicus*, Marshall and Orr (1952) obtained a curve relating reproductive rate and food concentration that had the same form as the curve produced from Green's results for *D. magna* (Fig. 3.16). However, attempts to establish a clear relation between reproductive rate and concentration of food organisms under natural conditions have failed (Edmondson, 1965; Edmondson, Comita and Anderson, 1962). This may be due either to the presence of adequate amounts of food in the water body or to some other predominant influences, probably mainly temperature.

It therefore appears that what data we have suggests that some relationship does exist between fecundity and reproductive rate, on the one hand, and temperature and trophic conditions, on the other hand, but that the dependence is not the same in all species of animals. More information and a further clarification of general patterns on this subject require careful systematic studies.

Chapter 4

METHODS FOR ESTIMATING THE PRODUCTION OF POPULATIONS WITHOUT CONTINUOUS RECRUITMENT

4.1 Boysen-Jensen Method

The reproduction of many aquatic animals takes place during rather short periods of time. Then a period of individual growth follows reproduction during which no new recruitment of young occurs and the numbers of that generation or cohort[1] decreases. Species populations with this type of reproductive pattern do not have a continuous recruitment of young. The production of such populations has been computed by many investigators who have either used the original method developed by Boysen-Jensen (1919) or have independently devised similar methods.

Boysen-Jensen determined the production of populations of the important species of benthic animals living in two bays of the Limfjord (Denmark). The small bivalves and polychaete worms were studied most intensively because these were important food organisms for plaice and eels and the production estimates were continued for a number of years (from 1910 to 1917). Every year, in spring and in autumn, he sampled with a bottom-sampler a series of stations more or less widely distributed over various areas in the bays. The annual production was calculated for the period from one spring (May) to the next in the following year. During this work Boysen-Jensen used certain terms to describe the various values he needed in his production calculations.

1. *Stock or initial biomass*—the numbers and weight of individuals in the spring samples.

2. *Residual stock or biomass*—the numbers and weight of individuals which remain by the spring of the following year.

3. *Consumption*—the numbers and weight of individuals which have been eaten (and so are no longer present) during the year.[2]

4. *Growth increment*—the increase in weight of all individuals during the course of the year, including the increment of those that perished during this period.

---

[1] The term "cohort" signifies the aggregate of individuals of a given species born at one time and which live together under identical conditions (Neess and Dugdale, 1959).

[2] What Boysen-Jensen calls consumption corresponds to the total loss by consumption and by other causes of death and henceforth is represented by E or by the index e (B_e, N_e).

5. *The appearance of the young or recruitment* ("*upgrowth*")—the numbers and weight of newly born young, this year's O-group.

6. *Annual production*—the sum of the growth increment and recruitment of young.

All these quantities were related to a unit surface area (1 m^2).

The biomass at the beginning and end of the year, as well as the new recruitment of young, was determined directly from the bottom samples. The magnitude for consumption (B_e) was taken to be the difference between the initial (N_1) and final (N_2) numbers, multiplied by the arithmetical mean of the initial (B_1/N_1) and final (B_2/N_2) mean individual weights of the groups of animals under consideration. Thus B_e is given by the equation[1]

$$B_e = (N_1 - N_2)\tfrac{1}{2}(B_1/N_1 + B_2/N_2) \tag{4.1}$$

where N_1 and N_2 are the numbers of individuals in the population at the beginning and end of the year and B_1 and B_2 are the biomass at the beginning and end of the year.

The growth increment is obtained from the sum of the consumption and the stock or biomass remaining at the end of the year, after subtracting the initial biomass. In other words, this quantity represents the production of the older generations (excluding the O-group) and may be expressed as P, where

$$P = B_e + B_2 - B_1.$$

Thus, Boysen-Jensen's annual production, that is the production of the whole population, is the sum of the increments for individuals of the older year-classes present at the beginning of the year and the increments of the newly born individuals of the new generation.

Production is easiest to compute in monocyclic species in which all the individuals of a population belong to a single year age-class and the new young appear only once a year at a time when few individuals remain which were born the previous spring. Thus, from one spring to the next, we are only concerned with decline in numbers and a simultaneous increase in the mean individual weight.

More frequently, as Boysen-Jensen points out, it is usual to deal with populations consisting of individuals of two or more age-classes. In order to compute the annual production under these circumstances, it is necessary to be able to distinguish the newly recruited individuals (O-group) from the older ones. Then the separately calculated productions of the young born during that year and of the other age-classes can be added together. Boysen-

---

[1] Here, and henceforth, the various authors' original symbols have been changed into the system of symbols which has been adopted for this book, in order to facilitate the comparison of different methods of computation.

Jensen was able to distinguish sufficiently well the new generation in the bivalves Solen, Abra, Corbula and Mya. He used data on the rate of growth in weight and made some assumptions about the degree to which the young were being consumed during the summer and the winter periods. According to his computations for the fjord which he studied, the annual production of the benthos during 1910–1915 varied from 42·1 to 72·1 g m^{-2} wet weight and the spring biomass from 22·3 to 60·8 g m^{-2}.

It is necessary to record that the computation of annual production, based on only two series of observations of the numbers and biomass of a population, can yield only a very approximate estimation of its real magnitudes, particularly in those species with rather short life-cycles whose reproductive periods are somewhat protracted or are repeated. Later Blegvad (1928), who worked at the same fishery laboratory, recognized the limited accuracy of Boysen-Jensen's calculations and also the fact that his estimates for production were too low.

Nevertheless, the basic principles of the computations used by Boysen-Jensen have received wide recognition and have been used and further developed by a number of investigators.

4.2 Production of Homotopic Benthic Animals

The work of Markosyan (1948), carried out in 1938–1939, is one of the first studies of the production of freshwater homotopic animals. It is devoted to the lake-dwelling Gammarus (G. lacustris) of Lake Sevan. Markosyan studied carefully the ecology and biological characteristics of this species, which occurs very numerously in the lake. He found that it had a two-year cycle of development and produced four broods during a rather protracted reproductive period (July–September) and, from seasonal changes in numbers and biomass of coexisting age-groups, he determined the annual production. In the Chara zone this came to 42·3 g m^{-2} whereas the mean annual biomass in the same zone was 20·6 g m^{-2}. Thus, according to Markosyan's data, the annual P/B coefficient for Gammarus lacustris in Lake Sevan was 2.

Later Bekman (1954) investigated another ecological form of the same species from a small eutrophic lake which is subject to oxygen depletion and situated on the flood plain of the Angara River. In this lake Gammarus had an annual life-cycle, a compressed period of reproduction and a number of other biological peculiarities. Production was calculated by the method of Kuznetsov (1948a, p. 84) from data obtained from four surveys. The sum of the loss of biomass of the parents and young classes came to 98 g m^{-2} and the annual P/B coefficient, calculated from the mean biomass, was about 3.

In a widely known work by Greze (1951) an example is given of the computation of production for the amphipod Pontoporeia affinis, based on detailed observations conducted throughout the year in two types of water in northern regions, in a river (1941) and in a lake which has a lower water temperature (the annual sums of degree-days were 1060 and 500 respectively). In the river the crustacean had an annual life-cycle of 14 months while in the lake it had a two-year cycle (27 months), with a short period of spring reproduction in both places. From this data Greze constructed curves of

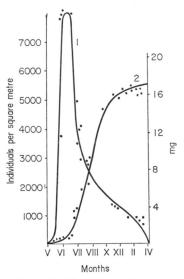

FIG. 4.1 Changes in the population density of one generation of *Pontoporeia* and its growth in the river.

1—density (individuals per square metre);
2—growth increment in mg (Greze, 1951).

numbers and growth rate for a year class throughout the whole of its existence (Fig. 4.1).

To calculate the production for the 15th of each month, the numbers N were read off from the curve. The number of individuals that died during the month was determined from the difference in numbers between any one month (N_2) and the previous month (N_1). Thus, for the river population, it was found from the curve that on July 15th there were 4800 individuals per square metre with a mean weight of 0·4 mg, whilst on 15th August the corresponding figures were 3700 and 2·2 mg. Consequently, during the month the numbers decreased by 1100 individuals per square metre.

For the weight of the individuals which died, Greze used the weight read off from the curve opposite the middle of the period under consideration

(1st August), which was 0·9 mg. Thus the biomass of individuals which died during the month was $1100 \times 0·9 = 990$ mg m^{-2}. The loss or elimination of biomass for all months of the year was determined in the same way and the sum of these was considered to be the annual production. For the river population the annual production came to 45·5 g m^{-2} whereas its average biomass was 13·10 g m^{-2}. The ratio of these two quantities gave the annual P/B coefficient, which came to 3·44.

In the northern lake which was studied, with its considerably lower water temperature, the annual P/B coefficient was 1·9.

Greze draws attention to the fact that the ratio of the total production of one generation to the original biomass of its parents (P/B_0) was almost the same in both populations, namely 3·2 and 3·4. He calls this index, which is expressed as a percentage, the "production constant".

The method of computing production used by Greze was also used by Bekman (1959, 1962) to calculate the production of a number of species of Lake Baikal amphipods. In Lake Baikal, *Micruropus kluki* inhabits sandy bottoms near the shore in Maloe More. The young of this species appear in massive numbers from June to August. The parental generation disappears towards the winter, after having lived a little more than a year. The annual production of the population (11·4 g m^{-2}) is 2·5 times both the average biomass (4·6 g m^{-2}) and the original biomass (4·5 g m^{-2}).

Comparable data are not given for the other amphipods; it is indicated only that a P/B coefficient close to 3 was also obtained for *Micruropus possolskii* and *Crypturopus inflatus*, both of which inhabit shallow water. Preliminary data for two populations of amphipods living at a depth of about 80 m indicated that they had a longer life-cycle and their productive capacity was 2–3 times less than that of the inhabitants of the shallower water.

In a later work, Bekman (1962) gives an account of the production of two species of endemic Baikal amphipods that inhabit shallow water of the eutrophic Posol'ski Sor, which is joined to Lake Baikal by a narrow channel; these are *Micruropus possolskii* Sow. and *Gmelinoides fasciatus* Stelb.

With an average yearly biomass of 5 g m^{-2}, the annual production of *M. possolskii* comes to 18 g m^{-2} ($P/B = 3·6$). For *G. fasciatus*, $P = 3·5$ and $B = 1·2$ g m^{-2} ($P/B = 2·9$). In this body of water both species have an annual life-cycle. There is also some information that for the population of *M. possolskii* in Lake Zagli-Nur $P/B = 3·4$ and $P = 23$ g m^{-2}, whereas for *G. fasciatus*, for the open coasts of Lake Baikal where its growth continues over two years, $P/B = 1·6$.

The works of Greze and Bekman show that, for homotopic animals such as amphipods, it is possible to calculate production when there is data on the numbers and individual growth of the animals throughout the whole duration of the life of one generation. In this very simple situation the initial and final biomass of the year-class is equal to zero and the production of a cohort balances the elimination.

For each period between two observations the elimination (E) is equal to

$$E = (N_2 - N_1)\bar{w}.$$

If there are n periods of observation during the whole duration of the existence of a cohort, the production of a cohort throughout its life is equal to

$$P = E_1 + E_2 + E_3 + \ldots + E_n.$$

If the cohort lives for several years, then its production must be obtained for the first, second and all subsequent years of its existence. The total production of the population in this case is the sum of the corresponding fractions of the production of each cohort.

A very informative investigation of the production of the amphipod *Hyalella azteca* in the rather small, shallow and eutrophic Sugar Loaf Lake (Michigan, USA) has been published by Cooper (1965). Very detailed weekly observations at four stations provided data on the age and size composition of the population. These data were supplemented by laboratory observations, from which was determined the length of the developmental stages at different temperatures, and the amount of growth during each moult. For the different stages the relationship between the stage of development and the total number of segments in the first and second antennae was determined, as well as the relationship between the number of segments of the antennae and the length of the head capsule.

For each weekly observation, the number of eggs, young (stages 1–3), juvenile (stages 4–5), newly mature and old individuals was determined. Knowing the duration of the developmental stages at any temperature, expressed in days (D), Cooper was able to calculate the size of the recruitment per day, which he took to be equal to $1/D \times N_0$, where N_0 is the initial number of the previous stage of development at the beginning of the period under consideration. He decided that use of the mean number for the previous stage during the time considered (a week) would give more accurate results than the initial number. Using this method of computation, the size of recruitment was calculated for each set of sampling data and each age-group, as well as the numbers of each stage which would be attained in a week in the absence of mortality. From the difference between the latter numbers and the actual observed numbers, the elimination was determined, or, in Cooper's terminology, the mortality of each of the stages and its changes throughout the time of observation. Cooper computed production as the product of mortality and mean biomass. Here mortality must be expressed as a fraction of the biomass during the time under consideration. Assembling the totals from his computations, he believed that in the lake he studied the mean magnitude of the production of the *Hyalella azteca* population came to 0·129 kg dry weight per hectare per day, and that the average biomass was 4·07 kg ha^{-1}. Thus the P/B coefficient per day was 0·032. If we postulate that the vegetative period lasts four months, we obtain an annual P/B coefficient of 3·8.

Cooper's work is of exceptional interest in many respects. In particular, he shows convincingly that the magnitude of the recruitment is proportional to $1/D$ only when the numerical abundance of the previous stage is stable, and he examines how a decrease in the numbers of females laying eggs affects the expected recruitment of young, using a specially developed formula. It is difficult to say what the real importance of this correction may be. Cooper introduces it somewhat strangely when discussing the computation of the number of eggs expected to complete development, apparently not realizing that it applies equally well to the calculation of recruitment for all other stages.

The method for computing production used by Cooper is very similar to the methods for determining production of populations with continuous recruitment described in Chapter 5.

The work of Kuznetsov (1941, 1948a, b, c) is of much interest for he studied in detail the biology of a number of littoral molluscs and several other invertebrates in the Barents Sea and part of the White Sea, and determined their production. In these studies the essential data on growth, production and so on were obtained from observations on animals kept in special cages set out in the places where they occurred naturally.

Kuznetsov calculated production as the loss of substance during the period concerned, to which he added the increment for the surviving individuals. The former quantity was calculated from the formula

$$P = [\tfrac{1}{2}(a_1 - a) + a]S$$

where a and a_1 are the average weights of an individual at the beginning and end of a period and S is the decrease in numbers during the period. Converting to our own symbols, $a = B_1/N_1$ and $S = (N_1 - N_2)$. By simple transformations, it is easy to demonstrate that Kuznetsov's formula is identical with that of 4.1.

To calculate the production of a whole population it is necessary to separate the individual year-classes and compute the production of each of them. The sum of the productions of these year-classes gives the production of the whole population.

As an example, we will take the calculation of the production of *Lacuna pallidula* from 11th November 1939 through to 25th May 1940 from the data given in Table 4.1. From the above formula the production of the winter and summer generations of 1939 were calculated to be 31·40 and 14·02 g m^{-2}. The increment in body weight of an individual belonging to these generations and surviving through to the second period of observations was determined. The increment of individuals of the winter generation was $30 \times (26·7 - 12·8) = 0·42$ g m^{-2}, while that of the summer generation was $680 \times (13·5 - 3·6) = 6·73$ g m^{-2}. The oldest animals do not grow during the winter and hence only the November biomass of the two oldest generations, which did not survive to the second period of observation, enters into the production, along with the biomass of the winter generation of 1940. In addition, it was found that the biomass of the eggs that were laid and died during the winter[1] came to

TABLE 4.1

Characteristics of the population of *Lacuna pallidula* living on *Fucus serratus* in the inlet of Dalne-Zelenetskii

	11.xi.1939			25.v.1940		
Generation	Number N_1 (ind m^{-2})	Biomass B_1 (g m^{-2})	B_1/N_1 (mg)	Number N_2 (ind m^{-2})	Biomass B_2 (g m^{-2})	B_2/N_2 (mg)
Winter 1938	80	7·00	87·5	—	—	—
Summer 1938	770	39·40	51·2	—	—	—
Winter 1939	1620	20·70	12·8	30	0·80	26·7
Summer 1939	2320	8·30	3·6	680	9·20	13·5
Winter 1940	—	—	—	280	0·40	1·4

20 g m^{-2}. Thus the production of the whole population during this time was $7·00 + 39·00 + 31·4 + 0·42 + 6·73 + 0·40 + 20·0 = 119·4$ g m^{-2}. In the same way, the production from spring to autumn was computed and this amounted to 170 g m^{-2}.

The other species, *Lacuna vincta*, differs from the previous one in that there is a planktonic larval stage in its life-history. This feature complicates the determination of the magnitude of loss (production) during the first stages of the life of a generation. In addition to the information on the fecundity of the females and numbers of groups of eggs liberated, it was necessary to have some information on the survival of the early stages, which was obtained from the observations on the cages. The production of *L. vincta* was estimated per square metre of area occupied by the seaweed *Laminaria*, where the molluscs congregated during reproduction in the spring. Annual production, calculated for four areas, came to 56, 151, 227 and 996 g m^{-2} wet weight. The ratio of

---

[1] We would like to comment that it is not completely clear whether the biomass of eggs laid should be included in the computation of production since their substance must have been partly included in the computed biomass of the older year-classes.

production (227 g m^{-2}) to the original biomass, at the time of reproduction, was about 5.

The annual production of *Margarita helicina* for the four areas varied from 15 to 2286 g m^{-2} and the P/B coefficient came to 3·4, 3·5, 6·3 and 15·1.

The possibility of applying the magnitudes of production so obtained on small individual areas to the whole littoral region depends on how representative were the areas selected, and on how accurately can be determined the density of animals living on shore vegetation and performing large-scale migrations. This seems to be the main difficulty in studying the production of the littoral fauna and the main source of potential error.

Kuznetsov's combination of field work with observations on enclosures have given valuable results.

4.3 The Production of Heterotopic Animals

The production of heterotopic species, such as chironomids, is more difficult to estimate than that of homotopic animals because the emergence of the imago of the former from the water body has to be taken into account.

As early as 1926, Lundbeck undertook to calculate the production of *Chironomus* larvae in the Plensk Lake. From data obtained in dredge samples, he established the period of maximal numbers and the time of the biomass maximum of larvae. Usually these two maxima do not coincide since large numbers of young larvae form a small biomass, and as the animals are growing the general biomass increases despite the fall in numbers. To the values of maximal biomass Lundbeck added the biomass of those larvae that died during the period from their peak number to their biomass maximum, plus the mean weight increment of larvae in the subsequent period, including also the weight increment of larvae eliminated in the periods between serial samples. The sum of these values was taken by Lundbeck to represent the annual production, which in one case was double and in another treble the maximal biomass.

The limitation of this method of production estimation lies in the fact that, when protracted periods intervene between observations during the birth of the young, an underestimate might be made of both the eliminated young and the increase in the mean weight increment of one larva, because the newly born keep replacing the eliminated individuals. At the time of imago emergence the growth increment may again be underestimated as a result of pupation and emergence of the larger individuals and also because two species are being investigated together. The newly born individuals of *Chironomus bathophilus* during the summer lowered the calculated weight increment of larval *Chironomus plumosus*.

Borutski (1939a, b), in his detailed investigation during 1935 and 1936 of the population dynamics and production of *Chironomus plumosus* of Lake Beloye in Kosino, was largely able to overcome this difficulty, and in 1938, prior to the publication of Borutski's paper, Yablonskaya (1947) adopted his method to investigate the production of *Chironomus plumosus* in two small lakes near Moscow called the Bol'shoe and the Maloe Medvezh'e.

The principle of Borutski's method consists of separating out and quantitatively estimating all the stages of development of every species from sufficiently frequent samples. Such analysis enables the duration of the life-cycle, the number of generations, the periods of hatching and the growth rates of the young to be determined. At permanent stations on Lake Beloye dredge samples were taken at eight depths ranging from 2 to 13 m three times a month in summer and twice a month during spring and autumn. In the subsequent treatment of the samples taken with a dredge sampler of his own construction, Borutski recorded the number of egg layings, larvae, pupating larvae and dead individuals. Every larva or pupa was weighed alive on a spring (torsion) balance. While the weights were being recorded, they were analysed into size classes with 5 mg intervals; dry weight was also determined.

To count the number of emerging insects and ovipositions, he placed special emergence cages at the sampling stations in the lake. These traps were examined every one or two days and the trapped insects were counted and weighed.

This extensive and thorough treatment of his data allowed Borutski to describe in great detail the distribution of the larvae within the lake, to determine the periods when mass emergence of the imagines occurred and when the new generation appeared. Corresponding to the three observed emergences and the life-span of one generation, three periods were noted, namely, from spring to 20th June 1935, from 20th June to 1st September 1935, and from 1st September 1935 to 20th June 1936. For each of these periods, Borutski established the initial total number of larvae at the time of peak numbers of young N_{max}, the total number of larvae and pupae found dead N_m, the number of emerged imagines N_i, and the number of larvae surviving at the end of the period N_r. All of his calculations were carried out without formulae or symbolic notations for the various values. By using the symbols adopted and employed above, we may write the basic equation which actually guided him in his calculation of the numerical balance of one generation during the investigated period as follows:

$$N_{max} - N_r = N_m + N_i + N_f.$$

The value N_f represents the number of "eaten" larvae and pupae which he estimated by difference, i.e.

$$N_f = N_{max} - (N_m + N_i + N_r).$$

Borutski did not use statistical methods for evaluating the authenticity of the values he obtained, but he examined many possible sources of error. In particular, he believed that it had not been possible to account for all the dead larvae by observation. It is also clear that the traps provided only approximate estimates of the emergences of the imagines and the number of ovipositions. Now that more sophisticated methods have been developed for determining the production, it has become clear that certain of Borutski's assumptions are without foundation. For instance, he believed that it was possible to compare directly the numbers of larvae and pupae found in the samples,

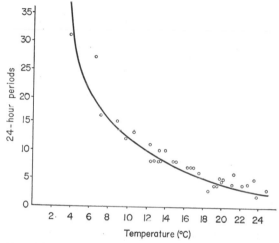

FIG. 4.2 Rate of decomposition of larval *Chironomus plumosus* at various temperatures (Yablonskaya, 1947).

tacitly supposing that the number of pupae found equalled the total number of pupated larvae. This assumption can only be made when the duration of pupal development exactly corresponds to the inter-sample interval. This misconception does not affect the basis of his computations expressed in the above equation, which gives the number of emerged imagines determined by the traps, and is not expressed in terms of pupated larvae. As Borutski himself emphasized, the number of emerged imagines was determined with a very low degree of accuracy and had to be corroborated by comparison with the total number of pupated larvae.

Having obtained detailed data on the distribution of larvae in the lake and on their size composition at each sampling date and every sampling station, he was able to convert his numerical values into units of weight with the help of the mean weights of larvae, pupae and imagines and so to express all items in the balance equation in terms of dry weight and for the lake as a whole. The weight of dead larvae was calculated from the body length and a previously established length–weight regression. In order to determine correctly the numbers of larvae that had died in the inter-sample intervals, it was necessary during the warm season, as shown by Yablonskaya (1947), to apply an appropriate correction for the rate of decomposition of the corpses. For this it is also essential to take the temperature of the bottom substrate as well as of the water layers just above, since the rate of decomposition of larval bodies increases with rise in temperature (Fig. 4.2).

TABLE 4.2

Decline in the individual dry weight of *Chironomus plumosus* during development

Developmental stage	According to Borutski (1939)		According to Yablonskaya (1947)	
	Dry weight		Dry weight	
	(mg)	(%)	(mg)	(%)
Larva prior to pupation	6·4	100·0	5·00	100·0
Pupa	5·5	85·9	4·18	83·6
Female imago	5·7	89·1	3·25	65·0
Male imago	3·4	53·1	2·66	53·2

Changes in the numbers and biomass at different stations are known to be due to the sometimes frequent migrations of larvae. Borutski overcame this difficulty by calculating the final production from values determined for the entire population (for the lake as a whole) and expressed in kg dry weight.

This computation may be expressed by the equation

$$P = B_m + B_i + B_f + B_d + (B_2 - B_1)$$

where B_m is the biomass of dead larvae and pupae, B_i is the biomass of emerged imagines, B_f is the biomass of consumed larvae and pupae, B_2 is the final and B_1 the initial biomass and B_d is the loss of substance during metamorphosis. This latter value was considered by Borutski and Yablonskaya, to be the weight loss of an individual emerging from pupation as an imago (see Table 4.2).

Borutski considered the value B_d, defined in this way, to represent the weight of organic matter lost in the water with the exuviae and other waste material from ecdysis; this justifies the inclusion of B_d as one of the items whose sum gives production. Note that this is only partly true, as the decrease in dry weight at metamorphosis is to a large extent the result of its metabolic demands.

The summarized data of Borutski (1939a, b) and Yablonskaya (1947) are presented in Tables 4.2 and 4.3, where they are expressed in grammes wet weight per square metre in order to permit comparison with the work of other authors.

TABLE 4.3

The annual production of *Chironomus plumosus* in three lakes, as determined by Borutski's method

Lake	Annual production P wet weight (g m^{-2})	Mean summer biomass B wet weight (g m^{-2})	P/B	Author
Beloye	34·70	14·63	2·4	Borutski
Bol'shoe Medvezh'e	106·30	35·80	3·0	Yablonskaya
Maloc Medvezh'c	76·20	28·80	2·6	Yablonskaya

Note. The proportion of *Chironomus* production eliminated by consumption was 26·5 per cent in Lake Beloye and 54 per cent in Lake Bol'shoe Medvezh'e.

TABLE 4.4

Calculation of the eliminated biomass of *Chironomus plumosus* in Lake Maloe Medvezh'e (Yablonskaya, 1947)

Period of observation	Development completed and emerged		Consumed by predators		Died between moults		Total loss (kg dry weight)	
	Nos. 10^6	Mean dry weight of 1 larva (mg)*	Nos. 10^6	Mean dry weight of 1 larva (mg)	Nos. 10^6	Mean dry weight of 1 larva (mg)†	Borutski	Boysen-Jensen
5.vi–6.vii	0·3	6·80	75·1	2·03	176·2	0·58	256	511
6.vii–20.ix	10·2	5·70	95·1	4·60	72·0	2·96	709	816
20.ix–16.iv	0·8	6·08	4·2	4·50	35·3	2·77	122	181
16.iv–17.vi	7·4	6·90	2·8	6·80	5·7	7·53	113	108
5.vi–17.vi	—	—	—	—	—	—	1200	1616

* Weight of larva prior to pupation.
† Weight obtained from length measurements of the dead larvae.

Borutski's method for calculating the production of *Chironomus plumosus* in Lake Beloye is applicable to those cases where the species being studied has a small number of readily distinguished generations with short periods of emergence and hatching of the young. For example, in the case of two generations per year, the calculations can be readily carried out by dividing the year into the following four periods: (1) from the onset of the imago spring emergence to the hatching of the spring generation of young; (2) to the beginning of the summer–autumn emergence of imagines; (3) to the hatching of the autumn generation of young; (4) to the beginning of the spring emergence of imagines. An important advantage of this method arises from the separate treatment of dead, consumed and emerged individuals, including the determination of their size and state of development, which makes it possible to define more precisely the weight of an animal at its elimination or at its emergence into the atmosphere, whereas Boysen-Jensen, Lundbeck and others made do with the mean individual weight of the population in their calculations. Sometimes the differences may be considerable, as is shown in Table 4.4, where Borutski's calculations are compared with those of Boysen-Jensen (last column of Table). As can be seen, Boysen-Jensen's values are 1·5 times greater than Borutski's.

It has already been mentioned that Borutski determined the weight of all the individuals in each sample—an accurate but time-consuming procedure. Satisfactory determinations of individual weight can be obtained for the various size groups from a previously established regression between weight and some linear dimension.

The developmental stages of chironomid larvae can be distinguished from the dimensions of their head capsules which show a sudden increase after every moult (Yablonskaya, 1947; Konstantinov, 1951)—Fig. 4.3.

According to Yablonskaya's data, a newly hatched larva of *Chironomus plumosus*, just beginning to feed, had a body length of 1·2 mm and a head capsule width of 0·15 mm. These dimensions were 2·2–3·2 mm and 0·27 mm respectively after the first moult, 5·5–9·5 mm and 0·50 after the second moult, and 9·7–28·0 mm and 0·99 mm after the third moult; after the third moult, the larvae still showed a considerable increase in size and weight. Thus head capsule measurements are not adequate to distinguish stage IV larvae among larvae belonging to different generations, and it is necessary also to measure their body length or to determine their live (wet) weight, or their dry weight. Providing the regression between length and weight is known for a given species, larval weight can be obtained from its body length (see Table 3.1, p. 34). This relationship can be established satisfactorily from determinations of the live (wet) or dry weights of larvae classified into size groups with class intervals of 1 or 2 mm (Yablonskaya, 1947).

Concurrently with the studies of Borutski and later Yablonskaya, Kirpichenko conducted detailed investigations on the benthos of two small lakes on the flood bed of the Dniepr (Lakes Ts'ganskoe and Podbornoe). Independently of other authors, he developed a method of calculating production that was based upon frequent sampling, at five-day intervals, of the numbers and biomass of the benthic fauna and of the chironomid larvae in

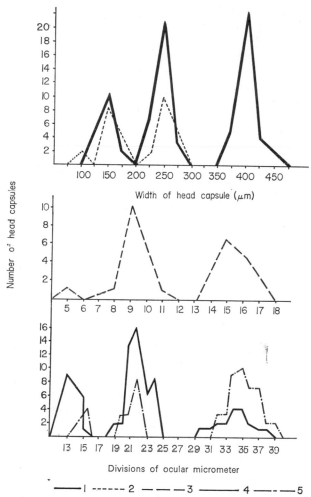

FIG. 4.3 Width of the larval head capsule in *Polypedilum gr. nubeculosum* (1), *Tanytarsus gr. mancus* (2), *Cricotopus gr. silvestris* (3), *Procladius* sp. (4), *Anatopynia gr. varia* (5) (Konstantinov, 1951).

particular (Kirpichenko, 1940). Production for the brief inter-sample period was taken to be the summed weight increments of larvae that survived and of larvae that died during the period. Recognizing that mortality could have taken place on any day between sample dates, Kirpichenko calculated the number of eliminated larvae, assuming a linear arithmetical decrease in numbers, and multiplied this by the daily mean weight increment of an individual larva. As Winberg and Koblents-Mishke (1966) point out, this method of calculating production is much the same as that given by equation 4.1.

Sokolova (1966, 1968) recently obtained some interesting results using Borutski's method to evaluate the production of many chironomid species of the Uchinsk reservoir. She expanded the application of his method to quantitatively collected early larval stages from the microbenthos, which was particularly important for the small species. The elimination of the first and second larval instars was determined from the difference between the calculated number of eggs laid and the number of larvae of a given species present in the samples of microbenthos. The number of eggs laid was established from the number of ovipositions detected in specially designed traps.

Prior to detailed investigations, the distribution of each species in the water body had been studied and this helped in selecting appropriate sampling stations. The emergence period for each species was established, also the time at which the young hatched, the number of generations, and the characteristics of the growth curve at different seasons. It was thus possible to establish from this data the best periods for estimating the production of the various generations for each species.

In all 17 species investigated the ratio of the production of one chironomid generation to its maximal biomass, usually attained shortly before emergence, ranged between 1·10 to 5·10 or, excluding the extreme values, between 1·3 and 4·4. The P/B coefficient, calculated from the annual mean rather than the maximal biomass, reached very high values (20·5 to 36·0) in species with a short life-cycle (Tanytarsini, *Polypedilum scalaenum*, *Cryptochironomus* and others). In species with a more protracted development and with only one generation per year (*Chironomus anthracinus*, *Psilotanypus limnicola*), the yearly P/B coefficient did not exceed 6·6 (from 2·9 to 6·6). Species with an incomplete second generation during the year (*Chironomus plumosus*, *Polypedilum nubeculosum*) showed intermediate values.

It appears from Sokolova's data that, in the Uchinsk reservoir which has not been fished for several years, 80–90 per cent of all the hatched larvae were taken as food by fish.

An original and apparently very serviceable method of calculating the production of a cohort of growing individuals has been proposed by Neess and Dugdale (1959) who have applied it to chironomid production.

The essentials of this method are illustrated in the graph shown in Fig. 4.4. Mean individual weight w_t, ranging from weight at hatching w_0 to weight at pupation w_p, is plotted on the horizontal axis and the corresponding numbers N_t are on the vertical axis.

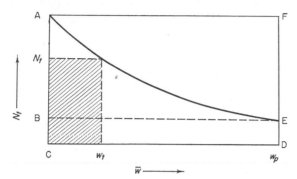

FIG. 4.4 Schematic graph relating numbers of larvae N_t and mean individual weight $\bar{w}$ (see text, Neess and Dugdale, 1959).

It is evident that for a mean weight of w_t and a corresponding number of N_t, the biomass is given by $w_t N_t$, which is represented on the graph by the rectangular area enclosed by the sides w_t and N_t.

Production for the period of time t, when the mean weight reaches the value w_t, is given by

$$P_t = \int_{w_0}^{w_t} N_t dw.$$

The authors follow Clarke *et al.* (1946) in terming this value as the "actual net[1] production". In Fig. 4.4, it is represented by the area ACDE.

From the same graph, Neess and Dugdale demonstrate the following terms.

Pupal biomass or "the actual net increase in standing crop" of Clarke *et al.* (1946) is given by the area BCDE.

"Directly recycled production" is represented by the area ABE and corresponds to what has been called elimination earlier in this book.

Potential net production, i.e. production in the absence of mortality ($N_0 w_p$), is given by the area ACDF.

In this way, a plot of numbers as a function of mean weight and a planimetric (for example) determination of the corresponding areas can provide an estimate of production and elimination. Neess and Dugdale applied their

---

[1] Here net production is contrasted with gross production, which is understood as the sum of the growth increment and metabolic loss. It is better to term the latter assimilation or incorporated food so that there is no need to use the term "net production", which appears to be equivalent to the term "production".

own method of computing production to previously published numbers and means weights of *Tanytarsus jucundus* (Walker) from Sugar Loaf Lake, where this species produces one generation per year (Anderson and Hooper, 1956) and obtained a value of 0·247. Earlier estimates by Anderson on the same data, using Boysen-Jensen's method, produced a value of 0·1745 (both values are expressed as ml volume of larvae per 0·25 m² per year).

Neess and Dugdale postulated that, under the ideal conditions of unchanging instantaneous growth k_q and mortality k_m rates and for given values of N_0, w_0 and w_p, production is dependent only on the ratio k_q/k_m.

When this ratio k_q/k_m is a constant, then

$$\lg w_t = \lg w_0 + (k_q/k_m) \lg (N_0/N_t).$$

For example, $\lg N_t$ is linearly related to $\lg w_t$, as is shown in Fig. 4.5. Neess and Dugdale demonstrated that this is true for the observed data on the numbers and mean weight of *Tanytarsus jucundus* (Walker) in Sugar Loaf Lake. In this example, production can be obtained not only planimetrically

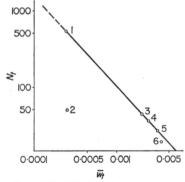

FIG. 4.5 Relationship between the numbers $\bar{N}_t$ and mean individual weight $\bar{w}_t$ of larva *Tanytarsus jucundus* in Sugar Loaf Lake. Logarithmic scales.

1 — August sample; 2 — September sample; 3 — December sample; 4 — January sample; 5 — March sample; 6—April sample (Neess and Dugdale, 1959).

from the graph $N_t = f(w_t)$ but also by calculation using the authors' formula which, however, is unlikely to have much practical application. A constant k_q/k_m ratio is likely to be found only in certain cases, as in insect larvae where growth proceeds at a near-constant rate up to pupation.

This method of estimating population production and elimination graphically by integrating the area of an empirical curve $N_t = f(w_t)$ for a cohort of growing individuals is an easy one for obtaining approximate values of these characteristics.

4.4 The Production of Populations of Planktonic Copepods

The same principle for estimating production, considered above for benthic animals, has also been applied for calculating the production of planktonic organisms with short periods of reproduction followed by a decline in population numbers of the species. Thus Yashnov (1940) used samples collected from the south-western region of the Barents Sea from June to April 1939 to calculate the production of *Calanus finmarchicus*, which was monocyclic in that area.

During the handling of the samples, the numbers of each developmental stage of the newly hatched generation were determined. From this data the percentage decrease in the observed population between sample dates was determined, so that the number that died at each stage could be calculated. It was then possible to calculate the number of individuals entering the next developmental stage as well as those remaining in the previous stage. The total number of individuals of each developmental stage that died before their metamorphosis into the next stage was determined from the summed data for all the periods of observation.

The production value which Yashnov terms "actual production"[1] was calculated as the sum of all eliminated individuals belonging to each developmental stage, plus those individuals that survived to the end of the period of study. To determine biomass, numbers were multiplied by their corresponding mean body weight.

Calanus had begun to reproduce before the start of the investigation and it was therefore assumed that the initial number of nauplii had exceeded the June crustacean numbers by 2·5 times. According to Yashnov's calculations, the annual production of *Calanus finmarchicus* came to 65 g m^{-2} and surpassed its maximal biomass by one and a half times or its mean biomass by 2·7 times.

For five years Kamshilov (1958) carried out monthly observations on the changes in numbers and age structure of a *Calanus finmarchicus* population in the coastal region of the Eastern Murman. He followed the seasonal change in numbers and biomass of Calanus and noted that its peak numbers occurred

---

[1] The term "actual production", used by other authors following Yashnov, has a meaning no different from that of "production" and, therefore, its use is unwarranted (see Chapter 1).

in mid-May. Following the maximum, a rapid drop in numbers was observed and, by the end of the year, there remained only a few individuals.

These data showed that numbers declined as an inverse exponential function of time, i.e. at a constant relative rate, which can be described by the formula

$$N_t = 695 \times 10^{-0.00814t}$$

where N_t represents numbers at time t, t is time in days, reckoned from the moment of maximal numbers, 695 was the level of maximal numbers per 1 m^3, and 0·00814 was the coefficient of elimination. With a daily coefficient of this order, the elimination $(1 - 10^{-0.00814t})$ is equal to a 0·019th part, or 1·9 per cent, of the population numbers. By multiplying the number of each stage by its mean individual weight for the corresponding period of the year, Kamshilov obtained a curve of population biomass showing a clear maximum in July. In order to compute the production of Calanus from the derived equation, the loss in population numbers was determined for short intervals of time (five days) and, from the biomass curve, the corresponding loss in biomass was found. The loss in biomass for a five-day period was obtained by multiplying the loss in numbers by the mean individual weight for the given five-day interval which, obviously, equals

$$\tfrac{1}{2}(B_1/N_1 + B_2/N_2).$$

The summed loss of biomass for the five-day intervals gave a value for the annual production for Calanus of 277 mg m^{-3}, which was 1·54 times greater than its maximal biomass $(P/B = 1·54)$.

These same principles of computation were employed by Timokhina (1964) for estimating the production of *Calanus finmarchicus* in the Norwegian Sea. In inshore waters and in the Atlantic Current, Calanus reproduces earlier— in May or sometimes in April—compared with animals from mixed waters or from the Eastern Icelandic Current, so that its production had to be determined separately for each water mass on the basis of samples collected monthly from April to November. The rate of biomass loss in young (I–III copepodid stages) and in adult (IV–VI stages) individuals was established separately. When any decline in the numbers of these two age-groups occurred simultaneously, this separation was not necessary.

The sum of the losses, plus the biomass of the crustaceans surviving to the end of observations, was taken to be the production for the period of the investigation. The production of Calanus ranged from 103 g m^{-2} for coastal waters to 10 g m^{-2} for Atlantic waters, whilst the mean for the whole area surveyed was 35·1 g m^{-2} in 1959 and 37·9 g m^{-2} in 1960.

Geinrikh (1956) tried out a preliminary calculation of the production of certain copepods in the Bering Sea. First, the annual cycle of reproduction and development of the crustaceans was studied from plankton samples. It was established that *Calanus cristatus, Calanus tonsus* and *Eucalanus bungii* were monocyclic in the western parts of the sea. Therefore, he could take the maximal biomass, which was observed in June, to be a major fraction of the production of these species. The loss of copepods in different stages of development and their weight increase when they metamorphosed into older stages, was determined by a comparison of copepod ages and numbers in regions where seasonal (phenological) variation occurred.

By adding to this maximal biomass the biomass of those copepods that died before reaching the maxima of earlier developmental stages plus the growth increments during metamorphosis to an older stage, Geinrikh was

TABLE 4.5

Annual production of copepods in the Bering Sea (Geinrikh, 1956) as grammes under 1 square metre

Species	Western region	Northern region
Calanus finmarchicus	—	5·2
Calanus tonsa	22	—
Calanus cristatus	26·5	—
Eucalanus bungii	51	1·6
Metridia pacifica	16	3·3
TOTAL	115·5	10·1

able to obtain some values for the production of these crustaceans. She calculated the production of the monocyclic species, *Calanus finmarchicus*, *Eucalanus bungii* and *Metridia pacifica*, from the northern part of the Bering Sea in a similar way (Table 4.5). As is indicated by the author herself, these calculations cannot claim to be very accurate and provide only a tentative idea about the level of production in the above species of Crustacea.

For his estimates of the production of *Calanus cristatus, Calanus plumchris* and *Eucalanus bungii* from the north-western regions of the Pacific Ocean, Mednikov (1960) also applied Boysen-Jensen's principles and determined production as the sum of lost biomass plus surviving biomass. He noted that the copepod numbers decreased sharply from May to September whereas their biomass increased, due to the development and growth of the copepodid stages during mid-summer. Such characteristic population dynamics provide suitable conditions for applying Boysen-Jensen's formula. However, it is possible that the computations were not equally accurate for all the species as Mednikov reports that the younger copepodid stages of *Eucalanus bungii* prevailed throughout the summer, which suggests that this species has a protracted reproductive period. In such circumstances it is not clear how the author manages to apply Boysen-Jensen's formula.

The production of the three indicated copepod species was, according to Mednikov, 41 g m^{-2}, and 60–70 g m^{-2}, when determined with more accuracy.

Yablonskaya and Lukonina (1962), determined the production of *Diaptomus salinus* in the Aral Sea. They, like the previous authors, considered that the annual production consisted of the sum of eliminated biomass B_e plus the biomass of individuals surviving to the end of the year B_r, i.e.

$$P = B_e + B_r.$$

The eliminated biomass was determined according to equation 4.1, but with slightly greater accuracy due to taking into account the biomass of the individuals hatching in the inter-sample period, i.e. in the time interval between the determinations of B_1 and B_2.

Consequently, equation 4.1 takes on the following form:

$$B_e = [N_1 - (N_2 - N_q)]\tfrac{1}{2}[B_1/N_1 + (B_2 - B_q)/(N_2 - N_q)] \qquad (4.2)$$

where N_q and B_q are the numbers and biomass of the young hatched during the period between observations.

Production estimates were based on samples from three plankton surveys (May, July and August) covering the whole sea from 1954 to 1957 and on year-round monthly observations at one sampling site. The number of newly hatched young was taken from the number of diaptomids present ranging in age from nauplii to the IIIrd copepodids, since development to the IIIrd copepodid stage requires barely a month. The more precise form of the formula for calculating elimination was not needed in this work since *Diaptomus salinus* is monocyclic in the Aral Sea.

The production values thus obtained for *Diaptomus salinus* in the Aral Sea ($P = 305$ mg m^{-3}; B (in spring) = 121 mg m^{-3}; $P/B = 2\cdot5$) must be underestimated since egg production is not included and the mean weights of males and females does not take into account their winter growth.

4.5 Recommendations for Estimating the Production of Populations without Continuous Recruitment

The methods available for estimating the production of species populations of a great variety of aquatic animals are all based upon common basic principles. However, in applying these common principles to particular cases, one encounters many characteristics that are peculiar to a given species or to a given set of circumstances. Therefore, before attempting an estimate of the production of any particular species, it is first necessary to have some general knowledge of the distribution of the species in its water body, its identification, the distinctive features of its various developmental stages and the age structure of its population. It is equally important to know the biology of the species, especially on the speed and nature of its reproduction. Such information will enable competent decisions to be made on where and how often to collect the quantitative samples which will be used to distinguish the changes in numbers and biomass of each age-group present in the population.

For populations permanently inhabiting a known biotope, samples can be collected from one appropriately chosen site. With species performing seasonal movements, it is necessary to take samples from a network of stations or along a transect traversing the area of the species' habitat. If information is needed about the total production of a species in the whole water body, samples must be taken throughout all the depth zones and all the bottom areas inhabited by the species in question, since under differing conditions its numbers, growth rate, the period and intensity of its reproduction as well as other biological features may be quite different. After the samples have been worked, results from stations lying in the same biotope can be averaged. In order to calculate values for the whole water body or for the total zone inhabited by a particular population, this averaged data must be weighted by the total volume of water occupied by each different biotope or depth zone.

Production can be estimated for various intervals of time—a decade, a month, a season, a year. Repeated series of samples are necessary and samples must be taken by a uniform method at the same points, transects or network stations. The intervals between consecutive observations are determined by the seasonal characteristics exhibited by the species in question. The shorter the life-cycle of the animal, the more frequent must sampling be in order to follow the changes occurring in its population. In studies on species with brief life-cycles, samples should be taken at intervals several times shorter than the duration of the whole cycle.

It is considered that with monthly observations good results can be obtained for species with annual cycles. As a rule the cumbersome year-round working procedure is not undertaken and in any event it is not strictly necessary in all cases. For species with a more protracted life-cycle, it is better to concentrate on the period of time when the breeding stages are developing, when numbers increase due to recruitment of young and when the cohorts of the young are growing intensively or are being eliminated. It is from such information, obtained from 3 to 5 periods of observation, that most of the calculations of annual production have been made for species with a life-cycle of a year or more (benthic and planktonic monocyclic forms). It appears that this is sufficient to obtain a necessary minimum of information in species with an annual cycle.

The productivity of animals living for several years under more or less stable environmental conditions can also be evaluated approximately from one well-planned, detailed sample collection made during or close to the reproductive period.

In all cases, sufficient samples should be collected to permit the determination of the errors associated with the numbers and biomass of different stages and any other results.

The collected field material is treated in the following manner. The specimens sorted from a sample are either measured or identified as various developmental stages. On this basis, the animals are divided into a certain number of age-groups or developmental stages which represent the composition of the population at a given period. The numbers of individuals belonging to each age-group are counted and their sum provides the total number of the species in that sample. The biomass of each stage or group is determined either by directly weighing all the sorted animals of each group or it can be calculated from a previously established size–weight or stage–weight relationship for individual animals. Addition of the biomasses of the separate groups gives the biomass of the species in question in the one sample, if necessary per cubic metre, per square metre and so on. From such a treatment of samples collected at different dates, information on the numbers and biomass of the population, as well as its various age-groups, is obtained and can be listed in a Table. Any changes in the number and age structure of a population can only be established by comparison of the results collected from two consecutive sampling dates. When the data available for estimating production appear consistent and regular, the figures in the Table may be used directly to characterize the population condition at a specific date. However, inevitably errors in determining numbers and biomass result in more or less pronounced deviations of these figures from the true values of corresponding parameters. Therefore considerably better results are usually achieved by constructing a time-dependent graph of numbers and individual weight for each generation and by using, for production estimates, not the direct empirical data related to a particular date but values read off from the curves. By such means, errors arising from deviations of isolated values are eliminated to a certain extent and irregular time intervals between samples can be replaced by periods of more uniform duration, etc. In order to fit the curve along the empirical points as well as possible, it is useful to know something about the general pattern of population dynamics and individual growth in the species being studied, as well as to employ the generally adopted methods for constructing empirical curves.

In order to estimate the production of invertebrates with a single reproductive period, it is convenient to start investigations when population numbers are maximal and this occurs at the peak period of reproduction. Subsequently numbers diminish and the relative importance of older age-groups and the mean individual weight in the population increases.

Elimination or loss in numbers N_e for the interval between observations $t_2 - t_1$ is obtained as the difference between initial numbers N_1 at time t_1 and the final numbers N_2 at time t_2, so that

$$N_e = N_1 - N_2. \tag{4.3}$$

The biomass of the eliminated individuals B_e equals the difference in numbers during the interval $t_2 - t_1$ multiplied by the mean weight of the eliminated individuals $\bar{w}_e$, thus

$$B_e = (N_1 - N_2)\bar{w}_e = N_e\bar{w}_e. \tag{4.4}$$

Where the time interval $t_2 - t_1$ is sufficiently small for the corresponding sections on the time-dependent curves of numbers and mean individual weight to be taken as being straight, then the mean individual weight can be determined simply as the arithmetic mean of the average weights at the beginning and end of the period concerned, i.e.

$$\bar{w}_e = \tfrac{1}{2}(\bar{w}_1 + \bar{w}_2) = \tfrac{1}{2}(B_1/N_1 + B_2/N_2). \tag{4.5}$$

In which case

$$B_e = N_e\tfrac{1}{2}(\bar{w}_1 + \bar{w}_2) = \tfrac{1}{2}(N_1 - N_2)\tfrac{1}{2}(B_1/N_1 + B_2/N_2). \tag{4.6}$$

Obviously, the biomass and the numbers of eliminated individuals for the whole of the period of investigation (whether for a year, a season, etc.) represents their summed values for the separate periods of observation.

Production for the period $t_2 - t_1$ can be computed in two ways. Considering first one of these, production is obtained as the summed eliminated biomass plus the difference between the final (B_2) and the initial (B_1) biomasses of the population:

$$P = B_e + (B_2 - B_1) = N_e\bar{w}_e + (N_2\bar{w}_2 - N_1\bar{w}_1). \tag{4.7}$$

As can be seen, production is equal to elimination, i.e. $P = B_e$, when the initial and final biomasses are the same ($B_2 - B_1 = 0$).

When, however, newly hatched young individuals continue to appear throughout the inter-sample period $t_2 - t_1$ and they are distinguishable from the rest of the population, then equation 4.1 should be used, so that

$$B_e = N_1 - (N_2 - N_q)\tfrac{1}{2}[B_1/N_1 + (B_2 - B_q)/(N_2 - N_q)]$$

where N_q and B_q represent the numbers and biomass of those young that were born during the period from t_1 to t_2.

Depending on the basic aims of the investigation, the final total of the calculated production is normally expressed as units of biomass. However, during the course of the study or for certain specific purposes, it is possible to state production in terms of numbers of individuals, which is obvious from the methods examined above and requires no special explanation.

Where not only the overall elimination is known but also how many individuals died naturally (N_m), were consumed by predators (N_f) and emerged as imagines (N_i), then the total number of individuals eliminated is given by

$$N_e = N_1 - N_2 = N_m + N_f + N_i \tag{4.8}$$

and their biomass equals

$$B_e = N_1\bar{w}_1 - N_2\bar{w}_2 = N_m\bar{w}_m + N_f\bar{w}_f + N_i\bar{w}_i. \tag{4.9}$$

In more detailed studies, the items on the right side of equations 4.8 and 4.9 may themselves represent earlier totalling, for example N_f may be the sum of invertebrates, eaten by fish and other predatory animals, and N_m, for chironomids, the sum of pupae and larvae that have perished.

The mean individual weight of an investigated developmental stage or size class can be determined by different methods and one should be chosen that is suitable for any particular organism or set of conditions.

Thus, the mean individual weight of stages either eaten by fish or that died can be determined from a previously established relationship between length–weight, head capsule size–weight, etc. Weight obtained in such a manner is often called the restored or reconstructed weight.

It is known that arthropods, when moulting, lose a certain quantity of organic material with their cast exuviae as well as in other ways. This lost matter is frequently disregarded which can obviously introduce, in some cases, a considerable error in the estimated overall amount of organic matter constituting the production of a species. To eliminate this error, the weight of substance lost during the metamorphosis from one developmental stage to the next should be added to its mean weight. Thus, the total biomass corresponding to the numbers of individuals consumed by predators will become $B_f = N_f(w_f + d)$, where d is the weight lost during metamorphosis to the next stage. However, it is not always advisable to make such a correction. For example, when attention is being concentrated on the question of how much matter and energy is being transferred via the food consumption of animals to the next trophic level, obviously the consumed biomass should be expressed without any correction for loss in organic material due to moulting.[1] It is therefore more accurate to sum the material losses at each moult (B_d) and consider the total value obtained as one of the items expressing elimination; thus,

$$B_e = B_m + B_f + B_i + B_d. \tag{4.10}$$

The second method of estimating production consists of summing, for a period of time, the weight increments of those individuals of the population that survived to the end of the period plus the weight increments of those that were eliminated during it,[2] i.e.

$$P = N_2 \Delta \bar{w}_2 + N_e \Delta \bar{w}_e \tag{4.11}$$

where $N_2 \Delta \bar{w}_2$ is the absolute growth increment for the period $t_2 - t_1$ of individuals surviving to the end of the period and $\Delta \bar{w}_e$ the mean absolute growth increment of those eliminated during the period. It is not difficult to find $\Delta \bar{w}_2$, as $\Delta \bar{w}_2 = \bar{w}_2 - \bar{w}_1$ (although strictly speaking this is true only if elimination is not dependent upon individual weight). If the sections of the individual growth and population number curves corresponding to the interval $t_2 - t_1$ can be considered linear, then it can be assumed that the mean weight of an eliminated individual is given by $\bar{w}_e = \frac{1}{2}(\bar{w}_1 + \bar{w}_2)$ and the corresponding weight increment by

$$\bar{w}_e - \bar{w}_1 = \Delta \bar{w}_e = \frac{1}{2}(\bar{w}_2 - \bar{w}_1).$$

Consequently, for this situation,

$$P = N_2(\bar{w}_2 - \bar{w}_1) + (N_1 - N_2)\frac{1}{2}(\bar{w}_2 - \bar{w}_1) = \frac{N_1 + N_2}{2}(\bar{w}_2 - \bar{w}_1), \tag{4.12}$$

---

[1] In chironomids and other insects this loss of organic material at emergence cannot be determined as the difference in dry weights of a pre-pupal larva and an imago. The reduced weight of the imago is also partly the result of respiratory loss during the pupal period—a source of loss not always considered when the magnitude of the correction is being determined (Borutski, 1939).

[2] Theoretically, it is easy to satisfy oneself that both methods of determining production arrive at identical results. Let n individuals, whose individual weights are $w_{e1}, w_{e2}, w_{e3} \ldots w_{en}$, be eliminated during the period $t_2 - t_1$, then $B_e = \sum_1^n w_e$ and formula 4.7 becomes

$$P = \sum_1^n w_e - N_1 \bar{w}_1 + N_2 \bar{w}_2.$$

The growth increments of the eliminated individuals is expressed as

$$N_e w_e = (w_{e1} - w_1) + (w_{e2} - w_2) + \ldots (w_{en} - w_n) = \sum_1^n w_e - n\bar{w}_1.$$

Recalling that $\Delta w_2 = \bar{w}_2 - \bar{w}_1$ and that $n = N_1 - N_2$, we can rewrite formula 4.11 as

$$P = N_2 \bar{w}_2 - N_2 \bar{w}_1 + \sum_1^n w_e - N_1 \bar{w}_1 + N_2 \bar{w}_2 = \sum_1^n w_e + N_2 \bar{w}_2 - N_1 \bar{w}_1.$$

so that the production for a sufficiently brief period of time can be calculated very simply as the arithmetic mean of the initial and final population numbers multiplied by the difference in mean individual weights at the beginning and end of that period.[1]

For longer inter-sample intervals, and when the corresponding sections of the growth and number curves cannot be considered linear, other methods for calculating mean weight and mean growth increment of the eliminated individuals have to be found. Precisely how to calculate Δw_e in such cases depends upon the nature of the growth and population number curves.

Amongst the various indices of productivity, or rate of production characteristic of a given species population under known conditions, the so-called P/B coefficient, the ratio of production to biomass, has particularly great significance. This P/B coefficient is fully realistic and has a constant significance only for stationary populations, that is populations with a constant age structure and biomass. For the non-stationary populations more usually encountered in nature, the P/B coefficient does not remain constant and represents only a certain mean value true for a defined period of time only. Obviously, multiplication of the P/B coefficient by biomass gives the production for the corresponding period of time.

Many investigators, and particularly those from the USA, term an analogous coefficient the "turnover rate" and its inverse form the "turnover time". This "turnover time" is nothing more nor less than the average duration of life of the species under a given set of conditions (L) and the "turnover rate" is $1/L$. The latter, like the P/B coefficient, has the time dimension to the power minus one. It is therefore not difficult to perceive that, under the ideal conditions of a stationary population (when $P/B = 1$), the population is entirely renewed within the period L. Therefore, P/B per unit time (for example 24 hours) equals $1/L$, provided L is expressed in units of 24 hours. Under field conditions this is also approximately true for average values of P/B and $1/L$ when sufficiently brief periods of time are being considered.

Naturally the "turnover rate" can be used to determine production whenever it is possible to assume realistically that a population is more or less in a steady state and when L has been found by correct methods and sufficiently accurately (not always easy to achieve). Methods for obtaining L will be considered later. It is sufficient to state here that L can be obtained as the

---

[1] Some authors, trying to find the production of eliminated individuals, calculated from the beginning the total number of individuals that died and the total number of days they survived, from which they obtained a 24-hourly elimination rate, $N_e/n = b$; this is multiplied by the mean 24-hourly growth increment which is $\gamma = w_2 - w_1/n$, where $n = t_2 - t_1$.

The sum of the arithmetic progression $(N_e - 0 \cdot 5b) + N_e 1 \cdot 5b) + \ldots + N_e - (n - 0 \cdot 5b)$ equals $\frac{1}{2} N_e n$ and the product of its multiplication by γ equals $(N_1 - N_2)\frac{1}{2}(w_2 - w_1)$. In other words, this needlessly complicated procedure arrives at the same formula as above (4.12).

quotient from dividing the numbers of the whole population or one of its developmental stages by its recruitment per unit time.

The P/B coefficient is computed in different ways and, as a result, may have various magnitudes—a fact which is not always realized. Of course, it can be computed only where production and biomass are expressed in the same units and for the same period of time.

The production value forming the numerator in the P/B coefficient may be related to any interval of time, 24 hours, a week, a month, a year, etc., and, correspondingly, one obtains a daily, weekly, monthly or yearly P/B coefficient. For the denominator it is common practice to use the mean biomass for the same period of time. However, often the annual P/B coefficient is required to relate the production to the maximal or minimal biomass of that year. In such cases, the coefficients should be defined as P/B_{max}, P/B_{min}. Other values of biomass have been used, for example, the spring biomass, the mean summer biomass, etc. There has even been a special modification of the annual P/B coefficient proposed, namely that relating production to the biomass of the spawning stock of the population. This coefficient, for which the term P/B_0 has been suggested to distinguish it from all the other P/B coefficients, cannot be less than one. It has been pointed out that this coefficient reflects the reproductive capacity of a species under the conditions of its own water body (Bekman and Menshutkin, 1964).

Which form of the P/B coefficient it is advisable to apply in each case depends on the biological features of the species being studied, the duration of its life-cycle, its environmental conditions and the aims of the research. Annual coefficients such as P/B_{max} and P/B_{min} and others may be useful, or even essential, for comparing the particular characteristics of populations of a variety of species or of the same species in different water bodies, but it should not be forgotten that to a considerable degree these coefficients are relative. Thus the magnitude of P/B_{max} depends greatly on the value for the maximal biomass, which could be higher or lower from chance causes. This relativity of the P/B coefficient can be demonstrated by an example. Let us assume that the P/B coefficient for a period of six months comes to 12. This may mean that production occurred during each of the six months and that the monthly coefficients were $12 : 6 = 2$. But it is quite possible that production took place during only two of the months with a monthly P/B coefficient of 6, whilst during the rest there was no production due, say, to low temperatures, although the level of biomass remained the same. Such considerations demonstrate that the most valid P/B coefficients are those calculated for relatively brief periods of time under relatively stable environmental conditions. These reflect unambiguously the population's average growth increment under known field conditions and can be compared with the metabolic intensity and individual growth rate of the species concerned.

P/B coefficients determined by observing these conditions in practice are equivalent to an averaged value of the relative growth increments of all the individuals present in the population during the period $t_2 - t_1$. To such P/B coefficients can be applied the ideas about the connection between growth rates and metabolic intensity which are discussed in the chapter on patterns of growth. Where the P/B coefficient has been calculated for a situation in which $B_2 - B_1 = 0$, that is, considering only the growth increments of the eliminated individuals, one may write $P/B = (t_2 - t_1)C_w$. Moreover, this is also true, to all intents and purposes, when the difference $(B_2 - B_1)$ is small. For example, if a P/B coefficient for an observation period of 10 days, calculated under such conditions, equals 2, then $C_w = 0.2(24 \text{ h})^{-1}$.

Wherever computed values of P/B coefficients are quoted, it ought to be clearly indicated exactly how they were obtained, to which period of the year and to which moment in the life-cycle they apply, to what conditions, particularly of temperature, the corresponding values of production and biomass relate.

One of the urgent immediate aims of production-orientated hydrobiological research is to try to characterize various species and environmental conditions by means of authentic coefficients. As more information in this field accumulates, it will be possible to replace the laborious painstaking work of determining production for various species populations with production estimates based upon the biomass together with established coefficients.

Chapter 5

METHODS FOR ESTIMATING THE PRODUCTION OF POPULATIONS WITH CONTINUOUS REPRODUCTION

5.1 Introduction

In contrast to those species in which a reproductive period is followed by one of decreasing population numbers, in polycyclic species the interaction between these two concurrent processes occurs throughout the whole season, namely, the increase in population numbers due to birth of young and the decrease in numbers from the death of animals. This complicates the estimation of production of those populations with continuous reproduction as the biology and normal patterns of growth and reproduction must be examined in more detail for each individual species.

In the first attempts made to evaluate the production of zooplanktonic species with relatively short life-spans and long continuous reproductive periods, coefficients for converting mass into production were used whose values were argued from first principles. Thus, Juday (1940) surmised that in Lake Mendota for an average year the life-cycle of zooplanktonic organisms was two weeks. From this he calculated that the annual "number of turn-overs" was 26 and obtained a figure for the annual plankton production by multiplying the mean biomass by this value. Later, in 1943, he used a coefficient with the value of 52 for estimates of production in Lake Mendota. Similar coefficients have been applied by Lindeman for production estimates in his well-known work of 1942, as well as by other American workers. It is not necessary to emphasize that only rough approximations of possible zooplankton production can be obtained in this way. Soviet research workers have also multiplied certain values of the P/B coefficient by the mean biomass of a given period of time in order to estimate the production of a population (Kozhov, 1950; Petrovich, 1954; Shcherbakov, 1956; Geinrikh, 1956).

A somewhat more detailed estimation of production was adopted for the first time by Meshkova (1952) in her well-known paper on the zooplanktonic production of Lake Sevan. She carried out detailed observations on the numbers and growth rates of separated size classes, according to the following procedure.

1. From an analysis of numerical changes of different size classes, the population number contributing to the annual production was calculated.

95

2. During sample analysis, all animals counted were sorted into size groups and a decision made as to which species population they belonged.

3. Changes in the population biomass were studied at short intervals (e.g. changes due to growth in weight, reproduction and loss in biomass).

4. The annual production was computed as the sum of biomass increments of all populations for the year. These summed results for Cladocera and Copepoda are summarized in Table 5.1.

TABLE 5.1

Biomass and production of daphnids and copepods in Lake Sevan (Meshkova, 1952)

	Biomass			
Organism	Maximum (mg m^{-3})	Annual mean (mg m^{-3})	Production (mg m^{-3})	P/B coefficient
Daphnia longispina sevanica eulimnetica	1061·04	342·72	2153·48	6·2
Acanthodiaptomus denticornis	66·62	29·4	76·25	2·6
Arcthodiaptomus bacillifer	122·76	58·77	148·16	2·5
Arcthodiaptomus spinosus var. *fadeevi*	349·91	180·92	367·91	2·0
Cyclops strenuus var. *sevani*	85·49	53·02	242·69	4·6

This method of estimating production involves laborious calculations. Moreover, Meshkova does not describe all the details of the calculation procedure, leaving a certain vagueness about how they were carried out.

Another method for estimating production of copepod crustaceans was proposed by Mednikov (1962). It was based on the assumption that mortality was constant throughout the development and that, when the temperature was the same, the speed of development of the various species differed very little. The production P is calculated as follows:

$$P = DN_0 \frac{(1 - e^{-kD_1})w_1}{D_1} + \frac{(e^{-kD_1} - e^{-k(D_1 + D_2)})w_2}{D_2}$$

$$+ \frac{(e^{-k(D_1 + D_2)} - e^{-k(D_1 + D_2 + D_3)})w_3}{D_3}$$

where N_0 is the initial number of a generation, determined as the product of the numbers of sexually mature females and their fecundity; D_1, D_2, D_3 are the developmental periods of nauplii, copepodids and the mean life-span of sexually mature individuals (in days) respectively; w_1, w_2, w_3 are the mean weights of individuals belonging to the same stages. The whole life-cycle

$D = D_1 + D_2 + D_3$ and k, the coefficient of elimination, is calculated from the formula

$$N_n = N_0 e^{-kD}.$$

For pelagic copepods, Mednikov makes the assumption that

$$(D_1 - D_2) = 125e^{-0.0833t}$$

where t is temperature (°C) and $D_1 : D_2 : D_3 = 1 : 1 : 2$.

This approach is only useful for preliminary estimates of the production of stationary copepod populations with a constant recruitment (i.e. when the numerical relationship between developmental stages and total numbers does not change during a period greater than D). Such stationary populations might occur among tropical or deep-water marine plankton. This approach is not applicable to other population states; that is why the method cannot be applied to freshwater zooplankton as attempted by V'yushkova (1965).

All subsequent attempts to calculate production have, to a great or lesser degree, used the ideas of Elster (1954). He considers that a change in the number of a species population is the resultant of two opposing processes, reproduction on the one hand and reduction in numbers due to consumption, death, etc., on the other.

Elster calculates the rate of increase in numbers due to reproduction by means of a "coefficient of renewal" which is the reciprocal of the period of development of the eggs in days, $E_k = 1/D_{ov}$ (Elster's notation). He decided to use time units of ten days and calculated "coefficients of renewal" for a series of ten-day periods. For the eggs of *Eudiaptomus gracilis*, during the summer months he found that E_k varied between 2 and 3, which represents an egg developmental period of 5 to 3·3 days, respectively.

The product of the renewal coefficient and total number of eggs in the population N_{ov} provides a value for the number of ovipositions during one time-unit, that is, it represents a value for recruitment. Mortality (destruction) or elimination E is obtained as the difference between this calculated value for recruitment $N_{ov}D_{ov}^{-1}$ and changes in number during the same period of time for which the renewal coefficient was calculated,

$$\text{(Elimination)} \quad E = N_{ov}D_{ov}^{-1} - (N_t - N_0).$$

Elster and his co-worker, Miss Eichhorn, who had investigated populations of *Diaptomus lacinatus*, *Acanthodiaptomus denticornis* and *Heterocope saliens* on the same basis, confined themselves to the calculation of renewal and elimination rates in terms of numbers of individuals and did not aim to estimate production, although they came close to this.

E

Elster's "coefficient of elimination" and other indices of the duration of a life stage or an age-size group are of very great significance for productivity research. In any comparison between numbers of different stages or age-size classes, it is fundamentally important that the duration of their existence be taken into account. For example, in a population with a constant recruitment rate, if stage A lasts 10 days and the subsequent stage B only 5 days, then at any one moment the population will contain half as many individuals in stage B as in stage A. It would therefore be mistaken to conclude that the number in stage B was fewer because of elimination. Despite this, such erroneous conclusions can be found even in recent studies; for example, by Hynes (1961) in his study on the production of aquatic insects in a small stream and by Crisp (1962) in his estimates of the production of aquatic corixid bugs. Such production estimates based on this faulty conception cannot be reliable.

It is useful always to remember that, with continuous recruitment and in the absence of selective mortality, the numerical ratios of the different stages or age-size classes in a population will be equal to the ratios of the reciprocals of their respective durations, that is,

$$N_1 : N_2 : N_3 = \ldots = 1/D_1 : 1/D_2 : 1/D_3 \ldots$$

This relationship should be taken into account, even if the period of the development of a particular stage or the length of duration of an age-size class has been established in nature only from the time between the peak numbers of one stage and that of the subsequent stage, i.e. by the time between the maximal abundance of nauplii and copepodids. This period of time is frequently surmised to be the duration of development of the earlier stage (of the nauplii, in our example). This is true only when the developmental periods of both stages are equal in length; ideally, where selective mortality or any other factor does not complicate the issue, the period between peak numbers of two consecutive stages represents half the developmental period of both stages or, in other words, the sum of half the developmental period of each of the two stages.

Stross and his co-workers (Stross, Neess and Hasler, 1961) used Elster's ideas on the replacement or renewal time to calculate "turnover time" for stationary populations of *Daphnia longispina* and *D. pulex* in the limed and un-limed parts of the Peter and Paul Lake:

$$\text{turnover time} = \frac{\text{number} \times \text{time interval}}{\text{number of recruits}}.$$

For any given time interval, production in terms of number of individuals is given by the number of recruits (newly born) which can be taken to be the total number of embryos in the population, provided the time interval adopted is equal to the duration of embryonic development.

The turnover time of *Daphnia longispina* in the limed part of the lake turned out to be 2·1 weeks, whereas that of *D. pulex* in the un-limed part was 4·6 weeks. The authors consider that such values of "turnover time" were indicative of conditions rather than characteristic of a species.

The turnover time of a population can also be calculated from the birth rate; for example, as in the studies by Hall (1964) for *Daphnia galeata mendotae* in Base Line Lake (State of Michigan) or by Wright (1965) for *D. galeata mendotae* and *D. schodleri* in Canyon Ferry Reservoir (State of Montana).

The rate of increase in animal numbers can be expressed by the following equation, provided that numerical changes due to mortality, emigration and immigration can be ignored:

$$N_t = N_0 e^{bt}$$

where b is the instantaneous birth rate, N_0 and N_t the initial and final numbers of animals and t the time interval in days.

The instantaneous birth rate can be calculated from "the rate of reproduction" B', provided the value of this is known (Edmondson, 1960), so that

$$B' = E/D$$

where B' is the number of eggs per day, E is the number of eggs per female, D is the embryonic period under prevailing conditions in days. The instantaneous birth rate is then $b = \ln(1+B')$.

The turnover time of a population can then be taken to be the reciprocal of the instantaneous birth rate, although strictly speaking this is true only for populations in a steady state.

In Base Line Lake the turnover time for the summer generations of *D. galeata mendotae* turned out to be very short, merely 4 days, which means that the turnover rate was very high in this population. In Canyon Ferry Reservoir it was 10 days for the same species and 6·7 days for *D. schodleri*.

It is possible that these differences in the turnover times for populations of the same species reflect temperature differences (Hall, 1964), although other factors may also be involved.

The method for calculating production just examined is valid only for populations in a steady state and with a mortality that is similar for the different age classes; the age structure of the population will therefore remain much the same. Where this is not so, the turnover time based upon population numbers will differ from the turnover time based upon biomass and this attractively simple method of calculation will be inapplicable.

A further development of Elster's idea can be found in the work of Petrovich, Shushkina and Pechen' (1964) on production of freshwater crustacean zooplankton. They used in their calculations not only recurrent ovipositions but also the reciprocal of the developmental duration of each stage against the previous one.

Hence, the production of each separate stage during a period of time t represents the difference between the number of individuals entering a given stage and the number entering the subsequent stage. For example, the production of copepod nauplii in terms of numbers can be obtained from the formula

$$P_n = (1/D_q)N_q t - (1/D_n)N_n t$$

where D_q and D_n are the developmental durations of the eggs and nauplii in days, N_q and N_n the numbers of eggs and nauplii at the beginning of the time interval t, in days.

Production in terms of numbers was converted to biomass units by Petrovich and co-authors (1964) by multiplying the total number of each stage by the mean individual weight and the production of the whole species population was estimated by summing the productions of each age-class. The production of the whole population of a species can therefore be obtained by substituting the mean weights of each stage in the formula

$$P = P_i + P_q + P_n + P_k$$
$$= \left(\frac{q'}{D_e} N_i x F + \frac{n'-q'}{D_q} N_q + \frac{k'-n'}{D_n} N_n + \frac{i'-k'}{D_k} N_k \right) t$$

where q', n', k' and i' are the mean weights of individual eggs, nauplii, copepodid and adults respectively; N_q, N_n, N_k and N_i the numbers of these stages at the beginning of the period of observation; D_q, D_n and D_k the duration in days of the development of eggs, nauplii and copepodids; D_e the interval of time in days between ovipositions; x the ratio of number of females to number of sexually mature individuals; F the number of eggs per oviposition.

Considering the ratios $(n'-q') : D_q$, $(k'-n') : D_n$ and $(i'-k') : D_k$, the numerator represents the difference in mean weights of two consecutive developmental stages and the denominator, the duration of development in days of a particular stage. These ratios provide an estimate of the growth rate of the various stages which is somewhat higher than the real mean growth rate since they are obtained from the mean rather than from the initial individual weight. It is, however, possible to use in the numerator the difference between the initial and final weight of any one stage (where the latter is also the initial weight of the next stage) and production can then be calculated from the following formula:

$$P = \left(\frac{q}{D_e} N_i xF + \frac{n-q}{D_q} N_q + \frac{k-n}{D_n} N_n + \frac{i-k}{D_k} N_k \right) t.$$

Here q, n, k and i are the initial weights of the respective stages. The ratios $(n-q) : D_q$, $(k-n) : D_n$ and so on provide an estimate of the growth rate of these stages which is much closer to the mean daily (24-hourly) growth increment.

To sum up, the production of each developmental stage per unit time is obtained from the product of its rate of individual weight increment and its numerical abundance in the water body. The production of the whole population represents the sum of the productions of its component stages.

Konstantinova (1961) used this approach to calculate the production of certain species of Cladocera. She considered that it was possible to represent the growth increment curve of these Branchiopoda by two intersecting straight lines covering two periods of development, that prior to and subsequent to attainment of sexual maturity, each period being in her opinion characterized by a constant daily growth increment. Therefore, according to Konstantinova's data, the relative daily growth increments (i.e. as a proportion of the body weight) of various species of Cladocera were 0·05 in *Simocephalus vetulus*, 0·04 in *Sida crystallina* and 0·02 in *Moina rectirostris* prior to sexual maturity, and 0·01, 0·02 and 0·009 respectively after attainment of sexual maturity. Knowing these values, and the numbers and weights of sexually mature and young animals, it is easy to calculate the production.

A somewhat similar method to that of Petrovich *et al.* (1964) described above, but rather more detailed, was employed by Lebedeva (1964) to estimate the production of *Daphnia longispina* in the Uchinsk Reservoir. The numbers of animals in the various size classes were determined at each given period. From special observations in cultures, growth increment curves were obtained which were used to convert field size classes to age classes.

Knowing the age structure, the numbers and biomass of a species population at any one moment, as well as the growth rate, it is possible to calculate the age structure, as well as numbers and biomass of that population after a certain period of time, assuming no elimination. The value obtained represents the so-called "calculated biomass". The difference between the calculated and the initial biomass represents the production of the species during the period involved, according to Lebedeva. In such calculations the production always balances elimination.

5.2 Graphical Method for Calculating Production

5.2.1 *The First Variant*

For any given period the production of a species population is represented by the sum of absolute growth increments of all the individuals in it. Every population consists of individuals of various ages, each with a different growth rate which is characteristic of their age. The specific growth increment usually declines with developmental age, whereas the absolute growth increment, which is given by the product of the specific growth increment and the individual weight, usually increases during the early developmental stages, reaches its maximum at a certain intermediate stage and decreases during the later ones. Therefore, in order to determine the production of a species, it is necessary to investigate both how the absolute growth increments vary with age under given conditions as well as the age structure of the population, that is, the numbers present of the various age-groups. Both of these functions of age can be obtained from empirical data and expressed in the form of curves which provide precisely the basic information needed for estimating production by the graphical method.

Where changes in individual weight can be described as a function of age, then the regression of absolute growth increment on weight can be obtained as a derivative of this function. Usually one is dealing with empirical data which does not provide such information and to construct a curve of absolute growth increment one must use the weight increment curve from a series of individuals.

We shall explain the construction of such curves and the procedure involved in estimating production with an example of a planktonic crustacean. Benthic forms are usually larger animals, producing fewer difficulties, and the computational techniques remain much the same.

TABLE 5.2

Values required for constructing a growth curve and age frequency distribution, taken from a population of *Cyclops* sp. in Lake Drivyaty in August 1964

Developmental stage	Initial weight $(10^{-3}$ mg)	Duration of development (days) D	Numbers of animals $(10^3$ m$^{-3})$ N	Mean numbers of each age class N/D
Eggs	0·236	2·6	19·9	7·1
Nauplii	0·236	10·2	11·2	1·1
Copepodites	8·0	12·8	7·9	0·6
Mature males and females	17·0	3·5	4·3	—

Note. The duration for each developmental stage is given for 18°C.

The weight increment curve of an individual can be built up from empirical data on the weights of different developmental stages. To plot such a curve for planktonic crustaceans, it is convenient to use the initial weights of stages

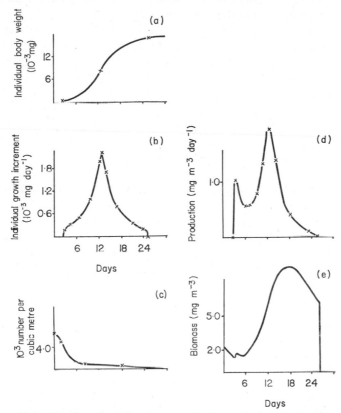

FIG. 5.1 A series of curves required to calculate production according to the first variant of the graphical method. The production of *Cyclops* sp. is given as an example.

(a)—growth of one individual; (b)—absolute growth increment; (c)—number of different age-classes in the species-population; (d)—production of all age-classes; (e)—biomass of all age-classes.

such as eggs, young and adults in Cladocera, for example, or eggs, nauplii, copepodids and adults in Copepoda (Table 5.2 and Fig. 5.1). Obviously, the initial weight of one stage is also the final weight of the preceding stage, disregarding the weight of exuviae and other losses on moulting.

The growth curve ends at the point of maximal individual weight for animals from a given population. If growth normally ceases upon onset of sexual maturity, then this point of maximal individual weight at which the growth curve ends is given by the initial weight of adult animals.

The initial weight of each stage is given by the youngest animals of that stage and should be plotted on the abscissa of the graph corresponding to the time of transition from the preceding stage.

Where growth continues even after the onset of sexual maturity, as in Cladocera, then the maximal adult weight must be plotted against that age ("maximal adult age") on the growth curve. However, it is rather difficult to determine the maximal life-span of adults under field conditions so that, where the number of very large individuals is low, their mean weight can be plotted at the point of mean life-span. For present purposes we can take it that the mean adult life-span is given by the formula

$$D = N_i/N_{io}^{11}$$

where N_i is the total number of adults and N_{io} is the number of youngest adults, i.e. those being recruited into adulthood per unit time.

Once a relationship between length and weight has been established for any species (Section 3.1), the minimal or initial weight of each developmental stage can be read off from its measured minimal length. Weight can also be determined directly or by other methods (see Chapter 3).

TABLE 5.3

Computation of the mean daily growth increment of different age classes (from the growth curve)

Time interval (days)	w_0 (10^{-3} mg)	w_1 (10^{-3} mg)	$w_1 - w_0 = \Delta w$	$t_1 - t_0 = \Delta t$ (days)	$\Delta w/\Delta t = P$ (10^{-3} mg day^{-1})	Mean time interval (days)
0	0·236	—	—	—	—	—
0–2·6	0·236	0·236	0	2·6	0	2·6
2·6–3·0	0·236	0·30	0·064	0·4	0·16	2·8
3·0–6·0	0·30	1·30	1·00	3·0	0·33	4·5
6·0–8·0	1·30	2·30	1·00	2·0	0·50	7·0
8·0–12·0	2·30	6·50	4·20	4·0	1·05	10·0
12·0–12·6	6·50	7·70	1·20	0·6	2·0	12·3
12·6–13·0	7·70	8·60	0·90	0·4	2·25	12·8
13·0–15·0	8·60	12·00	3·40	2·0	1·70	14·0
15·0–19·0	12·00	15·30	3·30	4·0	0·82	17·0
19·0–23·0	15·30	16·60	1·30	4·0	0·32	21·0
23·0–25·6	16·60	17·00	0·40	2·6	0·15	24·3
25·6–25·7	17·00	17·00	0	0·1	0	25·6

Since temperature influences the duration of development, any growth curves must take this factor into account. (The relationship between developmental period and temperature is considered in Section 3.2 as well as methods for calculating the durations of development at different temperatures.)

It is more difficult to advise on the influence of food abundance on the duration of development. One can only surmise that, for the development of possibly many populations under field conditions, temperature is a more significant factor than food quantity.

From the growth curve of an individual animal, constructed from empirical data, a curve of absolute growth increment related to age can be plotted (Fig. 5.1 (b); Table 5.3). If age is expressed in units of 24 hours (= days), this curve will describe a daily absolute growth increment. The translation of the simple growth curve to one of absolute growth increment can be carried out by one of the various methods of graphical differentiation. In the simplest method, the growth curve is subdivided into portions small enough to be equated to a straight line from which can be read off the initial (w_1) and final (w_2) weights, corresponding to the moments in time t_1 and t_2; the mean growth increment can then be computed for the period of observation from the formula

$$\Delta_w/\Delta_t = (w_2 - w_1)\,(t_2 - t_1)^{-1}.$$

When constructing the curve of absolute growth increment, each value of $\Delta w/\Delta t$ is plotted at the mid-point of the relevant time interval, that is, on the abscissa at $t_1 + \frac{1}{2}(t_2 + t_1)$. Moreover, the zero point should be placed at the end of embryonic growth (if it is assumed that the weight of an egg equals the initial weight of a newly hatched young individual.[1] From this curve relating the mean daily growth increment to age, it is possible to read off the growth increment of an individual at any moment of its development.

The construction of the third curve on age frequency is based on the following ideas. If a particular developmental stage lasts for several days, it will always consist of animals of different ages (copepodids may take five days to develop and will be between one and five days old). By dividing the total number of individuals of a particular stage by its developmental duration (see Table 5.2), a value is obtained which corresponds to the mean number of that stage per day of development. It is this value which is plotted on the ordinate of the age frequency curve (Fig. 5.1 (c)) at the mid-point of the

---

[1] In the graphs in various published works (Pechen' and Shushkina, 1964; Winberg *et al.* 1965), this point has been mistakenly placed at the mid-point of the egg developmental time; this error was noticed by V. A. Kostin.

stage's development. Such a curve will show the approximate number of each of the various possible ages within a stage at any particular moment. Usually, the younger the stage, the greater its number in the plankton and the shorter its duration; so that, as a general rule, the age frequency distribution is a curve falling off with age. However, a curve which rises with age is also possible during periods of reproductive inactivity.

It has already been pointed out that the product of the mean daily absolute growth increment and number in a particular age-group gives the daily production. These growth increments and numbers for each age-group can

TABLE 5.4

Computation of the production and biomass of a population of *Cyclops* sp.

Age (days)	P (mg day^{-1})	N (10^3 ind m^{-3})	P (mg m^{-3} day^{-1})	w (mg 10^{-3})	B (mg m^{-3})
0	—	10·0	—	0·236	2·36
1·3	—	7·7	—	0·236	1·82
2·6	0	5·4	0	0·236	1·27
3·4	0·26	4·0	1·04	0·4	1·60
6·0	0·42	1·3	0·55	1·3	1·69
7·0	0·50	1·1	0·55	1·8	1·98
9·0	0·78	1·0	0·78	3·1	3·10
11·0	1·38	0·95	1·31	5·0	4·75
12·8	2·25	0·85	1·92	8·0	6·80
14·0	1·70	0·80	1·36	10·5	8·40
18·0	0·62	0·65	0·40	14·8	9·62
23·0	0·21	0·45	0·10	16·6	7·47
25·6	0	0·35	0	17·0	5·95

$P = 16·61$ mg m^{-3} day^{-1}
$B = 140·42$ mg m^{-3}

be read off the relevant curve and multiplied to obtain production (Table 5.4); then, these values of daily productions of different age-groups can be plotted as a population production curve (Fig. 5.1 (d)). The area[1] under this curve (that is, the integral of the curve) provides the mean daily production of the members of the population which are growing; it does not include the growth increments of the adults or their egg production.

The production of eggs per unit time can be calculated from the rate at which eggs are laid and their biomass during that period of time.

---

[1] The area under the curve and the axis of the abscissa can be measured by various methods; for example, by planimeter, by counting the number of millimetre squares on graph paper or by weighing the cut-out curve on a balance. The accuracy required will be determined by the scales chosen.

$$P' = (q/D_e)N_i xF$$

where q is the weight of one egg, D_e the duration of an egg's development (in days), N_i the number of adults, x the ratio of females to the total number of individuals, and F the egg number per one oviposition. The total production for the whole population is then obtained by summing egg production with the growth production obtained from the production curve of different age-groups.

The total production of a population can be determined directly from the graphs described above if the growth of sexual products is incorporated into the curve of individual growth, i.e. if the body weight of an individual includes the weight of all eggs laid up to that moment.

Apart from the absolute values of production, it is particularly interesting to obtain also the ratio of production to mean biomass for the same period of observation, that is, the value of the P/B coefficient. It is, therefore, useful to determine as accurately as possible the biomass of the population, taking into account its age distribution. This can be done as follows.

To obtain the biomass of each age-group for the period in question, the numbers of each group, read off from the age frequency curve (Fig. 5.1 (c)), are multiplied by their mean individual weight, read off from the individual growth curve (Fig. 5.1 (a)). A curve of biomass (Fig. 5.1 (e)) can then be plotted and the measurement of the area under the curve (the integral) gives the biomass of the members of the population which are growing. The biomass of the non-growing individuals, such as adults which cease growth upon attaining sexual maturity, is obtained from the mean number and mean weight and can be added to that of the growing individuals.

The main difficulty encountered in using this method of estimating production is evaluating the life-span of adults under field conditions. This must be known in order to construct the growth curve and the age frequency curve of sexually mature adults which still continue to grow.

5.2.2 The Second Variant

Greze, in collaboration with Ten (Greze and Baldina, 1964), modified the graphical method of production estimation for population with uninterrupted recruitment. Their method arises from the same basic ideas, namely, that the growth curve of an individual reflects the process of production in a population and that the magnitude of the growth increment in an individual changes with age. Consequently, in any given population, if the numbers of individuals in each of its various age (weight) groups are known, it is possible to calculate the combined growth increments for a defined period of time, which represents the total population production.[1]

In order to calculate production, an individual growth curve must first be constructed for specified temperatures and other conditions; in this curve are included as growth increments any sexual products released from those individuals which are sexually mature (Fig. 5.2). As for production estimation, this growth curve should reflect the growth characteristics of both males and females, with possible significant differences in their growth rates; it must

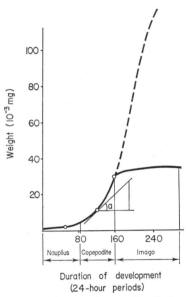

FIG. 5.2 The growth curve for *Cyclops kolensis* from Lake Baikal.

be plotted from averaged values which must also take into account the different proportions of males and females present.

For example, let us say that at a certain age t, females attain a weight of 0·025 mg and males 0·019 mg; the ratio of females to males is 2 : 1. The resulting mean weight of an individual aged t, averaged for this population,

---

[1] In 1963, Greze (*Zool. Zhurn.* **42**, 9) described a graphical method for estimating production, founded upon the same basic premises. In this paper, he proposed that a mean daily growth increment for an individual, which represented the average body size for the whole population, should be used to determine the production of that population. However, difficulties in defining these means, and errors arising in their determination by the graphical method, were such that this technique for calculating production was abandoned and the procedure outlined here was adopted.

will be $\dfrac{2(0 \cdot 025 + 0 \cdot 019)}{3} = 0 \cdot 023$ mg and this is the value to be plotted on the growth curve.

The daily growth increments of different developmental stages or animals of different sizes can be obtained from a series of points along the growth curve. At selected points on the curve, lines are drawn at a tangent and the daily growth increment is given by the tangent of angle a (Fig. 5.2). To determine

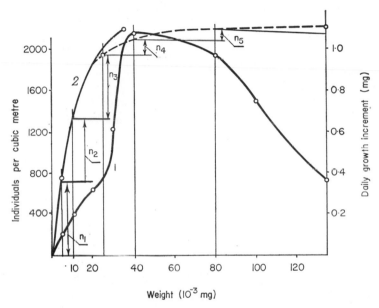

FIG. 5.3 The daily growth increment (1) and density (2) of a population of *Cyclops kolensis* as a function of individual weight.

this, it is necessary to read off correctly from the vertical and horizontal axes of the graph the weight increase in milligrammes (grammes) for the corresponding number of days. Dividing the former value by the latter provides the daily weight increment for the chosen portion of the graph. These growth increments, determined in this way, should be tabulated and plotted in a graph relating growth increment to weight (Fig. 5.3, curve 1).

The third graph needed for estimating production must reflect the size frequency in the population. The data for this graph can come from field observations of the population, that is, from samples analysed into numbers of different developmental stages or size classes. This data should be plotted cumulatively so that the highest point corresponds to the total number of individuals in the population sample. Thus, in our example of the *Cyclops kolensis* population in Lake Baikal (data from Mazepova, 1963), the number of nauplii was 750, of copepodids 1200, and of adults 240 individuals per cubic metre. The total number of individuals came to 2190 per cubic metre. When plotting curve 2 (Fig. 5.3) corresponding to these numbers, it was necessary to bear in mind that, from the growth curve, nauplii belong to the size class 5×10^{-3} mg or less and therefore the naupliar number 750 should be plotted at a point opposite 5×10^{-3}. Similarly, the growth curve indicates that copepodids have a weight not exceeding 22×10^{-3} mg. Therefore the summed numbers of copepodids plus nauplii, 1950 individuals per cubic metre, is plotted along the abscissa at a point opposite the weight, 22×10^{-3} mg. The total number of copepodids, including adults, is plotted opposite the maximal weight of about 35×10^{-3} mg, as is illustrated in the solidly drawn part of curve 2 (Fig. 5.3).

However, if we consider egg production to represent a further increment in individual weight, in *Cyclops kolensis* the maximal weight, together with the sexual products ($= 0\cdot1$ mg), comes to $0\cdot135$ mg. This is demonstrated in the dotted part of curve 2 which represents the sexually mature copepods which have produced various quantities of eggs and so belong to various weight categories. The dotted part of curve 2 has to serve for estimating production; the solid part of the curve can be used for biomass determinations.

Having combined the frequency of various weight classes and their corresponding daily growth increments in one graph, we can proceed to calculate the production. The graph must be subdivided into vertical sections each corresponding to a weight category. The width of these vertical sections should be determined by the nature of the changes in curves 1 and 2; where the curves show large changes the sections should be narrow, but they can be wider where the changes are small.

From this graph the mean individual daily growth increment (curve 1, Fig. 5.3) and the number (curve 2, Fig. 5.3) can be determined for each weight category and tabulated (Table 5.5). Production is obtained by multiplying these figures for each weight group and summing the products, as is shown in the right-hand column of Table 5.5.

As can be seen from the Table and graph, the number of individuals in the first weight group (less than 5×10^{-3} mg) is 750, in the second group (5×10^{-3} to 10×10^{-3} mg) it is 575 ($= 1325 - 750$), and so on. The corresponding mean individual growth increments (from curve 1, Fig. 5.3) are $0\cdot07 \times 10^{-3}$, $0\cdot15 \times 10^{-3}$ mg and so on. The products of these two values are $52\cdot5 \times 10^{-3}$, $86\cdot2 \times 10^{-3}$ mg, etc. which, totalled for the whole population of *Cyclops kolensis* per cubic metre, comes to $545\cdot7 \times 10^{-3}$ mg per day.

TABLE 5.5

Calculation of the production of *Cyclops kolensis*

Weight class (10^{-3} mg)	Number of individuals per cubic metre	Individual daily growth increment (10^{-3} mg)	Daily production (10^{-3} mg m^{-3})
0–5	750	0·07	52·5
6–10	575	0·15	86·2
10–25	625	0·29	181·2
25–40	130	0·87	113·1
40–80	100	1·06	106·0
80–135	10	0·67	6·7
TOTAL	2190	—	545·7

In this way the total production is estimated according to the simple formula

$$P = \sum_{i=1}^{k} N_i \bar{C}_{w_i}.$$

Such values of daily production, expressed in absolute weight units, can be related to the biomass of the producing population, where the biomass is readily determined from the numbers and mean weights of different weight groups, as is shown in Table 5.6. This relationship or P/B coefficient is an index of the relative growth increment of the population. In our example it is 545 : 23462 = 0·023.

Such daily P/B coefficients and daily absolute growth increments for populations can be obtained for longer or shorter periods of time, depending upon the basic information used to construct the curves for individual growth

TABLE 5.6

Calculation of the mean annual biomass of *Cyclops kolensis*

Weight class (10^{-3} mg)	Number of individuals per cubic metre N	Mean individual weight w (10^{-3} mg)	Biomass (10^{-3} mg m^{-3}) Nw
0–5	750	2·5	1875
5–10	575	7·5	4312
10–20	520	15·0	7800
20–35	345	26·5	9475
TOTAL	2190	—	23462

and weight composition of the population. If the information about the composition of the population has been obtained from more than one set of samples, then the intensity of production can be characterized for the period of the investigation and only very tentatively can these production indices be used for extrapolation for longer periods of time. Since it is essential to have more reliable data about production processes for longer periods of time (months, a year), it is desirable that observations on numbers and population composition be carried out systematically. Such observations will provide numerical data on the different stages or weight groups throughout a period of time. To calculate production for the period of time in question, average numbers for each group must be established, by graphical integration, and these used to construct a curve of mean size frequency in the population. Such a procedure enables one to regard the calculated values of the daily production and P/B coefficients as sufficiently representative mean values for the whole period of time under consideration; they can then be used to calculate the absolute production for that period of time.

Naturally, in considering the limits of such periods of time to be used for estimating production, variation in temperature conditions during the period must be considered and consequently also the growth rates of the organisms being investigated. When these periods are rather long, a curve of size frequency as well as of individual growth must be constructed for each one, taking into account the mean temperature of the relevant growth increment. It may be necessary to introduce temperature corrections based on the relationship between growth rate and temperature which has already been discussed.

Dividing the annual cycle of observations into four seasons—each with a known average temperature and known population composition—will probably provide calculated values of production which in practice are sufficiently reliable.

This assertion is not meant to imply that the method outlined here is not sufficiently accurate to require further development. What is badly needed is a more precise characterization of the growth curves of various taxonomic groups, whether parabolic, S-shaped or in some other form. We need to accumulate more experimental data enabling us to typify such curves.

The procedure described here was developed mainly for production estimates in organisms with prolonged periods of intensive reproduction and complex population structure, such as Cladocera and Copepoda. There is no theoretical reason why the method cannot be applied to other organisms with limited reproductive periods or with relatively simple population structures of a few generations. In such cases the population production can be calculated on the basis of its size composition at different periods and the characteristic growth increments of its different size groups. However, our experience in applying this method to such populations is so far insufficient. Production estimates obtained in this way should be compared with others calculated on the basis of population decrease. Where organisms reproduce two or three times a year and it is difficult to apply other methods for estimating population, it might be preferable to try the above procedure. One of our aims in future research will be to compare the various methods of production estimation and find out which is the best.

5.3 The Physiological Approach to Estimating the Production of Species Populations

The patterns of animal growth described in Chapter 3 form the basis for calculating production which is the sum of the growth increments of all the individuals in a population. Growth increments can also be calculated, with some approximations, from known values of respiratory intensity and the coefficient of utilization of assimilated food for growth (Winberg, 1966, 1967). For convenience we shall call this procedure for estimating production the physiological method. The advantages of this method are that approximate values for production of a given population can be obtained knowing only the numbers n_i and individual weights w_1 of the different age-groups, provided that the metabolic loss T and the coefficient of utilization of assimilated food for growth K_2, characteristic for the population, have been established.

Where the relationship between metabolism and individual weight is known, this can also be used in the physiological method for production estimation; this relationship is usually expressed by the equation

$$Q = Mw^{a/b} \tag{5.1}$$

where Q is the rate of oxygen consumption of an individual (ml h^{-1}), w is the individual weight (g), M is the coefficient representing the metabolic rate of an animal weighing 1 g and a/b is a constant.

From the rate of oxygen consumption, an approximate estimate of the metabolic loss can be made. Knowing the calorific equivalent of the animal's body weight, it is possible to calculate how much body substance has been spent on metabolic needs by an individual per unit time (metabolic loss T) using the equation

$$T = T_1 w^{a/b} \tag{5.2}$$

where T is the metabolic loss of an individual weighing w per unit time expressed in the same units as w, and T_1 is a constant representing the metabolic loss of an individual weighing one unit ($w = 1$).

The unknown quantity, production or growth increment P, is related to T via K_2, i.e.

$$K_2 = P/(P+T) \tag{5.3}$$

where P is the growth increment of an animal per unit time, expressed in the same units as T and K_2 is the coefficient of utilization of assimilated food energy for growth.

From equation 5.3,

$$P = TK_2/(1-K_2). \tag{5.3a}$$

Substituting in 5.3a the value of T taken from 5.2 we obtain

$$P = T_1 w^{a/b} K_2/(1-K_2) = Nw^{a/b} \tag{5.4}$$

where $N = T_1 K_2/(1-K_2)$.

Equation 5.4 represents the differential form of the equation for parabolic growth; in its integral form it can be written

$$w_t = [N(1-a/b)t + w_0^{1-a/b}]^{b/(b-a)} \tag{5.5}$$

where w_t is the weight of the animal at moment t and w_0 is the initial weight, for example that of an egg.

According to equation 5.4, the specific growth rate or relative growth increment C_w is expressed as[1]

$$C_w = Nw^{-(1-\frac{w}{a}/b)}. \tag{5.6}$$

Therefore the production of the whole population consisting of animals of i-ages will be

$$P = \sum_{i=0}^{i=D_m} C_{w_i} n_i w_i \tag{5.7}$$

where D_m is the duration of the development of the oldest age-group, n_i is the number of animals having an age i and w_i is their mean weight.

Equations 5.4, 5.5 and 5.6 can be used to determine the production of only those animals whose growth curve is nearly parabolic.

As mentioned in Section 3.1, where growth is parabolic, the coefficient of utilization of assimilated food for growth K_2 can be taken to be constant. In any other type of growth K_2 is not constant and production cannot be calculated from the above formulae; if the growth can be described by the

---

[1] Note that $P = \Delta w/\Delta t$. At a sufficiently small unit of time $P = dw/dt$ and then $Cw = (1/w)(dw/dt) = P/w$.

equations for S-shaped growth, then it may be determined on the basis of data given in Section 3.1. Consequently, one of the first requirements of the physiological method is to establish the type of growth involved.

Below is given an example of the application of this method to certain freshwater and marine copepods.

Judging from data in the literature and from our own results, it appears that copepod growth is more adequately described by a parabolic curve than by an S-shaped one (Figs. 5.4, 5.5 and 5.6), particularly if the growth increment due to the production of eggs is included in the curve (Figs. 5.7 and 5.8). Assuming, therefore, that copepod growth is described as a parabola, the pro-

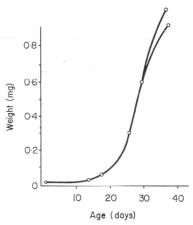

FIG. 5.4 The growth of *Calanus helgolandicus* (Petipa, 1965).

duction of Copepoda of any kind can be calculated with the help of equations 5.6 and 5.7.

We will take a Black Sea copepod, *Acartia clausi*, as an example for calculating production by this method. Data on the numbers and weights of different age stages were obtained from the work of Greze and Baldina (1964), who calculated the production of *Acartia clausi* for four seasons— spring, summer, autumn and winter—using the graphical method (second variant). From this same original data for *Acartia clausi*, the production was calculated by the physiological method, although only for the summer season. It was assumed that the calorific value C of the copepods was about 600 cal g^{-1} wet weight and that the oxycalorific equivalent $K_c = 4.86$ cal ml^{-1} oxygen. Therefore, if Q is expressed in ml of oxygen per hour, the metabolism of one individual, expressed in calories per 24 hours, will be given by equation 5.2, so that

$$Q \times 24 \times 4.86 = 116.64 \, Mw^{a/b}. \tag{5.2a}$$

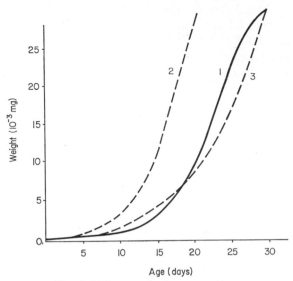

FIG. 5.5 The growth of *Acartia clausi*.

1—curve from the empirical data of Petipa (1965); 2—a theoretical curve of parabolic growth where $M = 0.437$ and $K_2 = 0.2$ (Petipa, 1965); 3—a theoretical curve of parabolic growth where $M = 0.300$ and $K_2 = 0.2$.

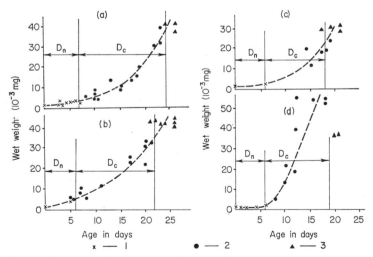

FIG. 5.6 Growth of *Macrocyclops albidus* in different concentrations of food (Klekowski and Shushkina, 1966).

1—nauplii; 2—copepodites; 3—adults; (a)—0.1 g m^{-3}; (b)—1.0 g m^{-3}; (c)—5.0 g m^{-3}; (d)—10.0 g m^{-3}.

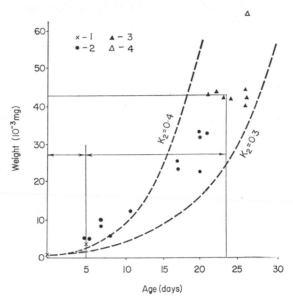

FIG. 5.7 Growth of *Macrocyclops albidus*, where $t = 21°C$ and the concentration of infusorian food was 1 g m^{-3} (Klekowski and Shushkina, 1968).

1—nauplii; 2—copepodites; 3—adults; 4—growth increment of an adult copepod due to the development of the first clutch of eggs. (1 and 2—growth curves calculated from equation 3.5 with $K_2 = 0.4$ or 0.3 and $Q = 0.165w^{0.8}$).

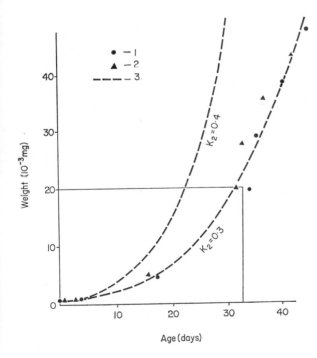

FIG. 5.8 Growth of individuals, including egg production, from a cyclopoid population from Belorussian lakes (averages for the season May to October) (Shushkina, 1965).

1 and 2—empirical data on cyclopoid growth from Lake Myastro (1) and Lake Batorin (2); 3—theoretical growth curves from $Q = 0.165w^{0.8}$ and $K_2 = 0.4$ or 0.3.

The metabolic loss per 24 hours is given by

$$\frac{116\cdot64}{600}\,Mw^{a/b} = 0\cdot195\,Mw^{a/b}. \tag{5.2b}$$

According to the data of many investigators, the value of the exponent a/b is close to $0\cdot8$ although the value of M, the coefficient of proportionality, can vary within rather wide limits. For example, Winberg (1956) cites for crustaceans at 20°C a value for M of $0\cdot165$ ml oxygen per gramme per hour. Petipa, investigating the metabolism of the Black Sea $A.$ $clausi$, obtained a value for M of $0\cdot288$ ml at 20°C and with $K_2 = 0\cdot2$.

In the literature (Table 5.7 and Fig. 5.9) values for K_2 range within the

TABLE 5.7

Published values of K_2

Author	K_2	Species or group	Comments
Ivlev (1939)	0·4	Carp	
Ivlev (1939)	0·6	*Silurus* larvae	Maximal value K_2
Ivlev (1939)	0·4–0·5	Pike fry	
Ivlev (1938)	0·45–0·5	*Paramecium caudatum*	
Tangl and Farkas (1903a, b)	0·6	Fish, chicken, frog, silkworm	Obtained during embryonic development
Tangl, Farkas and Mituchi (1908)	0·6		
Vasil'eva (1959)	0·45	*Daphnia pulex*	K_2 for a laboratory culture
Shpet and Kozadaeva (1963)	0·5	Carp	
Galkovskaya (1963)	0·5–0·6	*Brachionus calyciflorus*	K_2 for a growing population
Sushchenya (1964)	0·25	*Artemia salina*	
Winberg (1962)	0·5	Populations of aquatic animals	K_2 for growing populations
Winberg (1964)	0·6	Invertebrates	Maximal value of K_2
Petipa (1965)	0·2	*Acartia clausi*	Mean value of K_2 for the population
Petipa (1965)	0·4	*Calanus helgolandicus*	
Petipa *et al.* (1966)	0·4	*Penilia avirostris*	K_2 for young females
Petipa *et al.* (1966)	0·25	*Podon*	
Petipa *et al.* (1966)	0·25	*Evadne*	K_2 for females
Klekowski and Shushkina (1966)	0·5–0·6	*Macrocyclops albidus*	K_2 for growing crustaceans
Khmeleva (in print)	0·4	*Artemia salina*	
Ostapenya *et al.* (1968)	0·4	*Daphnia magna*	K_2 for an age differentiated population
Ostapenya *et al.* (1968)	0·4	*Daphnia pulex*	
Ostapenya *et al.* (1968)	0·25	*Moina rectirostris*	

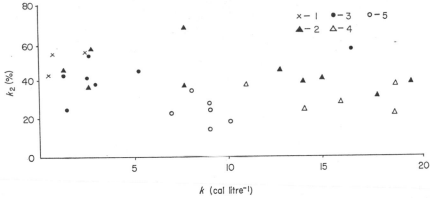

FIG. 5.9 Values for K_2, obtained by various authors at different food concentrations, k. 1—predatory zooplankton from three lakes (Winberg *et al.*, 1965); 2—*Brachionus calyciflorus* (Galkovskaya, 1963); 3—*Daphnia magna*; 4—*Daphnia pulex*; 5—*Moina rectirostris* (3, 4 and 5 from Ostapenya *et al.*, 1968).

limits of $0·6 > K_2 > 0·2$ and the mean value for K_2 can be taken to be about 0·3 for natural populations during the vegetative season (or season of growth). If we accept the following two sets of values for M and K_2 for *Acartia clausi*, based on data from two authors we have

(1) $M = 0·268$; $K_2 = 0·2$ (data of Petipa, 1965).
(2) $M = 0·165$; $K_2 = 0·3$ (averaged data for crustaceans, Winberg, 1965).
The following values for N were calculated from the equation 5.4:

$$(1) \quad N = 0·195 \times 0·288 \, \frac{0·2}{1 - 0·2} = 0·0141$$

$$(2) \quad N = 0·105 \times 0·165 \, \frac{0·3}{1 - 0·3} = 0·0139.$$

It therefore appears that very similar results were obtained for the intensity of production calculated from equations 5.4 and 5.6, first using values for metabolism Q and K_2 obtained by Petipa and then averaged values for crustaceans cited by Winberg. Details for calculating C_w and P for the whole population of *Acartia clausi*, employing the physiological method, are given in Table 5.8.

The production estimates for *Acartia clausi*, determined by three methods, happen to agree fairly well in terms of daily total production (see Table 5.9), despite the fact that the relative influence of the various age-groups is different in the three methods of calculation. Since *Acartia* stops growing on attaining sexual maturity, the adult mean weight was plotted on the growth curve at

TABLE 5.8

Calculation of the production of *Acartia clausi* by means of the physiological method and using the following values:
$C_w = Nw^{-0.2}$, $N = 0.014$, $P = \Sigma C_{w_i} n_i w_i$, $C = 0.6$ cal mg^{-1} wet weight, and (1) $M = 0.165$ and $K_2 = 0.3$; (2) $M = 0.29$ and $K_2 = 0.2$

Developmental stage	w (10^{-1} g)	lg w	lg w	-0.2 lg w	$w^{-0.2}$	C_w	n numbers	B (10^{-3} mg)	P (mg day^{-1})	Daily P/B
Nauplii	0·8	$\bar{7}$·903	-6·097	1·2119	16·6	0·23	730	584	0·136	0·233
Copepodites	6·0	$\bar{6}$·778	-5·222	1·044	11·1	0·16	406	2436	0·377	0·155
Adults	38·0	$\bar{5}$·580	-4·420	0·884	7·6	0·11	173	6400	0·685	0·107
TOTAL	—	—	—	—	—	—	—	$\left.\begin{array}{c}9420 \\ 9700*\end{array}\right\}$	1·198	0·124

* Total plus eggs.

TABLE 5.9

A comparison of three methods for determining the P/B coefficients for a Black Sea population of *Acartia clausi* (P in mg m^{-3})

Developmental stage	B (mg m^{-3})	B Percentage of whole	Graphical method								Physiological method			
			Variant I				Variant II							
			P	Percentage of whole	P/B		P	Percentage of whole	P/B		P	Percentage of whole	P/B	
			Daily		Daily	Monthly	Daily		Daily	Monthly	Daily		Daily	Monthly
Nauplii	0·58	6	0·15	15	0·26	7·7	0·26	21	0·44	13·2	0·14	11	0·23	7·0
Copepodites	2·44	25	0·64	62	0·26	7·8	0·80	66	0·33	9·8	0·38	31	0·16	4·6
Adults	6·40	66	0·24	23	0·04	1·1	0·15	13	0·02	0·7	0·68	57	0·11	3·2
Eggs	0·30	3	—	—	—	—	—	—	—	—	1·20	—	—	—
TOTAL	9·72	—	1·03	—	0·11	3·2	1·21	—	0·12	3·7	—	—	0·12	3·7

the time when adult life began (i.e. in our case the weight, $w_i = 38 \times 10^{-3}$ mg, should be plotted opposite 28 days and not 65 days (Fig. 5.10).

The example we have given demonstrates the possibility of using the physiological method. The more thoroughly studied are the metabolic rate, the type of growth and other original values, the more reliable will be the estimates it provides. Its main interest lies in the fact that the physiological

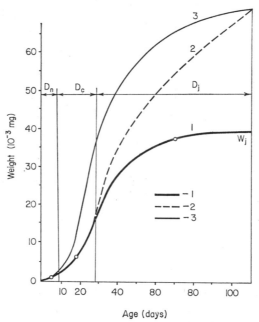

FIG. 5.10 Growth of *Acartia clausi* (Greze, 1963).
1—Greze's curve, excluding egg production; 2—the same curve, including egg production; 3—the corrected curve.

constants it utilizes appear to characterize not individual species but rather two ecologically similar animals.

The physiological method is useful for providing approximate production estimates for those animals whose growth rate at different ages cannot be determined experimentally but for whom, together with information on numbers and weight of different age-groups, the weight-specific metabolic intensity, the calorific values and the coefficient of utilization of assimilated food for growth are known.

For approximate production estimates of planktonic crustaceans, it is possible to assume that Q (ml of oxygen g^{-1} h^{-1} at 20°C) $= 0.165\ w^{0.8}$, if w is expressed in grammes of wet weight, and that $K_2 = 0.2 - 0.4$.

5.4 The Production of Planktonic Rotatoria

The combination of great reproductive powers with relatively brief life-cycles implies that rotifer production plays an important rôle in the conversion of matter in bodies of water.

However, most current methods of estimating production cannot be applied to Rotifera. These have been developed for species with distinguishable developmental stages and relatively long life-spans whereas, in rotifers, it is virtually impossible to determine size and age categories or to discriminate between sexually mature females without eggs and senile ones, etc. (Edmondson, 1965).

Edmondson (1960) was the first worker to use Elster's method (Elster, 1954, 1955) to calculate the natural reproductive rate B of egg-bearing planktonic rotifers, i.e. to determine the mean number of eggs laid per day by one female, which is given by

$$B' = E/D$$

where E is the mean number of eggs per female and D the duration of embryonic development in days.

Provided that the total population numbers show relatively little change during a brief period of time, it is possible to calculate the population production per unit time from the product of the mean numbers and reproductive rate B'. Therefore, the population production for time t expressed in numbers of individuals is given by the formula

$$P = B't\tfrac{1}{2}(N_0 + N_t) \tag{5.8}$$

where N_0 and N_t are the initial and final numbers of rotifers.

Without mortality and in the absence of factors limiting the increase in population numbers, after one day, instead of one individual (a female), $(B'+1)$ individuals will be present after two days and $(B'+1)^2$ individuals, etc. Therefore, in this example individual numbers are the exponential function of time, so that $N_t = N_0 e^{bt}$ where $b = \ln(B'+1)$ and is the rate of instantaneous increase of numbers. In fact, changes in population numbers depend not only on numbers but also on mortality. If we make the assumption that mortality per unit time is proportional to the number of individuals present, this can be indicated by a negative mortality constant, $-\alpha$, with a negative sign; changes in population numbers can then be expressed by the equation

$$N_t = N_0 e^{(b-\alpha)t}.$$

From this formula α can be calculated, provided that the values for b, N_t and N_0 are known for the period t. Knowing α and substituting in $\alpha = \ln(1-M)$, we can determine what proportion of the rotifers present are eliminated daily (M'). So now, having the mean number of rotifers present ($\overline{N}$) for the period in question, we can obtain elimination from $N_e = \overline{N}M't$ and production from $P = N_e + (N_t - N_0)$.

Clearly, errors can occur during the determination of the values of B' or the original parameters, E and D. For instance, it is very difficult to determine E when there is a daily rhythm in ovipositing (Edmondson, 1965; Galkovskaya, 1965). Moreover, during the collection of samples, some of the most widely distributed species of rotifers, such as *Keratella cochlearis*, *Ascomorpha* sp., *Chromogaster* sp. and *Polyarthra* spp. (Galkovskaya, 1965), might be lost through the bolting silk. Assuming that mainly the smallest, that is the youngest, organisms of each species are lost through the bolting silk, the value of B' could be greatly overestimated.

The method proposed by Tonolli (Edmondson, 1960, 1965) for determining the duration of embryonic development should be regarded with caution. According to this procedure the embryonic period is determined by the decrease in egg numbers in living samples. The time taken for the original number of eggs to disappear entirely from the sample is considered to be the duration of the embryonic period under the given conditions. Since the embryonic period commonly corresponds to the time between ovipositions, new eggs could have been laid during the period of observation, which is why the application of Tonolli's method could lead to an erroneous prolongation of the duration of embryonic development. There is almost no experimental data on the duration of embryonic development.

A less laborious and more general method for estimating production is proposed, based upon the inverse relationship of the relative daily individual growth increment to generation time (Galkovskaya, 1965), where the generation (or doubling) time is taken to be the period from the hatching of the parent rotifers to the hatching of their offspring. And the calculation can be carried out by the formula

$$P = D_2^{-1}t\tfrac{1}{2}(N_0 + N_t) \qquad (5.9)$$

where N_0 and N_t are the initial and final biomass or number of the species involved, t is the duration of the observation in days, D_2 the time taken for doubling and P the production during time t in terms of numbers (or biomass). To convert this to units of weight, the average size of an individual must be established for each period of time, from which the mean weight can be calculated.

This was the method used to estimate the production of rotifers in Lake Naroch for the period May to October 1961. The value of D_2 was determined for several species of planktonic Rotifera by culturing them individually in lake water at known temperatures (Table 5.10).

TABLE 5.10

The relative daily growth increment $(1/D_2)$ for several species of Rotifera

Species	$t°C$	$1/D_2$
Asplanchna priodonta	21	0·33
Keratella cochlearis	15	0·15
Kellicottia longispina	15	0·15
Kellicottia longispina	18	0·20
Euchlanis dilatata	19	0·25
Euchlanis dilatata	20	0·27
Synchaeta grandis	19	0·25
Anuraeopsis fissa	19	0·33
Epiphanes senta	20	0·25
Ploesoma hudsoni	19	0·28
Polyarthra spp.	15	0·25

Samples of sedimentary plankton, collected every two or three days by both vertical and horizontal hauls, provided a basis for estimation of production. The values obtained for rotifer production, calculated for each month observed, are summarized in Table 5.11.

TABLE 5.11

The biomass, production and mean daily P/B coefficients for planktonic rotifers in Lake Naroch from May to October 1961

Period of observation	B (mg litre^{-1}) Asplanchna	Other rotifers	P (mg litre^{-1}) Asplanchna	Other rotifers	Daily P/B coefficients Asplanchna	Other rotifers
May (15.v)	0·229	0·179	0·396	0·504	0·115	0·18
June	0·129	0·227	0·505	0·803	0·17	0·12
July	0·068*	0·120	0·088	0·636	0·26*	0·16
August	0·147	0·141	0·783	1·216	0·20	0·28
September	0·003	0·103	0·015	0·534	0·18	0·17
October (to 10.x)	0	0·131	0	0·190	—	0·14
15.v to 10.x	0·089	0·150	1·787	3·883	0·16	0·17

* The biomass values and the P/B coefficients have been calculated for the period 26th to 31st July.

In two species, *Keratella cochlearis* and *Kellicottia longispina*, production was estimated both by Edmondson's method and from the doubling time. The mean daily P/B coefficient for *Keratella cochlearis* was 0·14, obtained by Edmondson's method, and 0·18 from the doubling time; the corresponding results for *Kellicottia* were 0·08 and 0·14.

It would appear that the physiological method (Section 5.3) will be particularly useful for determining rotifer production. This approach is based upon direct measurement of the metabolic rate and efficiency of conversion of food energy into growth and production is calculated from the formula

$$P = TK_2(1-K_2) \tag{5.10}$$

where T is the metabolic loss in units of weight (energy) per unit time and K_2 is the coefficient of assimilated food used for growth.

In this method it is sufficient to know the levels of T and K_2 to obtain an estimate of production.

The oxygen consumption of four species of planktonic rotifers was determined in lake water. Table 5.12 presents the values obtained in relative units per unit time.

TABLE 5.12

Respiratory rates and metabolic loss in rotifers

Species	w $(10^{-3}\,\mathrm{cal\,ind^{-1}})$	Respiratory rate		T $(\mathrm{day^{-1}})$
		$(10^{-5}\,\mathrm{mg\,ind^{-1}\,h^{-1}})$	$(10^{-3}\,\mathrm{cal\,ind^{-1}\,day^{-1}})$	
Brachionus				
calyciflorus	0·75	0·17	0·14	0·18
Epiphanes senta	1·50	0·35	0·29	0·19
Filinia longiseta	0·11	0·026	0·021	0·19
Keratella				
cochlearis	0·094	0·032	0·027	0·28

The oxygen consumption of the rotifer *Brachionus calyciflorus* was measured at different concentrations of the alga, *Scenedesmus obliquus*. It can be seen from Table 5.13 that the rate of oxygen consumption is influenced considerably by the algal concentration.

Simultaneously, experiments on rates of growth and reproduction of other rotifers were carried out at the same concentrations of *Scenedesmus obliquus*. Table 5.14 gives the mean values of the relative food ration R/w, the assimilability $1/U$, and the coefficients of utilization of consumed (K_1) and assimilated (K_2) food for growth at different levels of available food. As can be

TABLE 5.13

The metabolic rate of *Brachionus calyciflorus* in different concentrations of food
(*Scenedesmus obliquus*)

Concentration of algal food	Metabolic rates		T, metabolic loss
cells 10^6 ml^{-1}	(10^{-3} mg O_2 mg^{-1} h^{-1})	(10^{-3} cal ind day^{-1})	(day^{-1})
0	0·80	0·13	0·17
0·2	3·30	0·55	0·73
0·5	2·45	0·41	0·55
0·8	3·10	0·52	0·70
1·0	4·80	0·81	1·08
2·0	2·45	0·41	0·55
5·0	2·10	0·35	0·47

seen, K_2 varies between 0·33 and 0·53 in a ten-fold increase in food concentration, but not in any regular way.

From Table 5.14 the average value of K_2 is 0·41, so that from equation 5.10 $P = 0·70T$. If T is expressed as daily weight loss, then P has the same value as the daily P/B coefficient in the example.

The measured rates of oxygen consumption for the four species of rotifers (*Brachionus calyciflorus, Epiphanes senta, Filinia longiseta, Keratella cochlearis*) provide the possibility of calculating values of T and so, from the formula $P = 0·70T$, values of daily production can be obtained, which are 0·13, 0·13, 0·13 and 0·19 respectively for the above species.

For these calculations metabolic rates at 20°C were used, which is the characteristic temperature for the summer months (July to August). For this

TABLE 5.14

Some basic indices relating to the utilization of food for growth in *Brachionus calyciflorus*

Biomass of algal food (mg litre^{-1})	Relative food ration (R/w)	Assimilability (U^{-1})	K_1	K_2
5	1·02	0·75	0·24	0·33
10	1·25	0·78	0·20	0·53
30	1·50	0·70	0·20	0·40
70	1·45	0·70	0·20	0·42
100	5·21	0·48	0·19	0·39
300	10·23	0·25	0·17	0·51

period of the year the mean daily P/B coefficients, calculated by doubling time, ranged between 0·19 and 0·23, and by Edmondson's method, between 0·13 and 0·17. So, all three methods for estimating production have yielded very similar results.

These results enable us to make some suggestions about the potential significance of rotifer production. A daily P/B coefficient of 0·15 corresponds to a monthly one of 4·5 and a daily coefficient of 0·20 to a monthly one of 6·0 and so on. In Lake Naroch the P/B coefficient of planktonic rotifers for the month of August was 6·8. This is a value which is two or three times greater than any monthly P/B coefficient obtained by Pechen' and Shushkina (1964) for planktonic crustacean species living in the same lake. What this means is that with a biomass two or three times smaller the production of the rotifer population was equivalent to that of the planktonic crustacea.

We are now in the position to be able to make tentative determinations of rotifer production, although all the proposed methods of calculation still require verification and refinement. More data needs to be accumulated on reproductive, growth and metabolic rates of various rotifers under different food and other conditions. Such investigations are urgently required.

5.5 Recommendations on Methods for Determining Production of Populations with Continuous Reproduction

Calculations of the production of animal populations with continuous recruitment are based on the same principles as those for populations without recruitment. But the feature common to all methods of production estimation in populations with continuous reproduction is the far greater difficulty in tracing changes in the mean weight and numbers of individuals in populations in which various age-groups are present simultaneously. To this difficulty can be attributed the imperfections of the methods we have examined so far for computing the production of such animals. It is impossible at present to recommend any one method, as was done in Section 4.5, since the method chosen largely depends on the nature of the investigation being undertaken (for example, whether it consists of an intensive or extensive series of observations). For an extensive study, aimed at obtaining some estimate of zooplankton production, the physiological method may be appropriate. It is true that it is a complicated matter to establish the physiological characteristics K_2 and T which make this method laborious. Nevertheless, it is suitable for evaluating orders of magnitude sufficiently accurately for extensive studies.

The choice of method depends also on the peculiarities of the life-cycle of the animal being studied. The method of Mednikov, examined earlier, can be applied only to stable populations with continuous recruitment—which is a rare enough occurrence. The calculation of production by turnover time, so widely used by research workers abroad, is based upon a uniform mortality rate for all age-groups in the population; otherwise, the turnover time for the population numbers would differ from the turnover time for its biomass. Thus, this method of calculation, so attractively simple at first glance, turns out to have some drawbacks. The graphical method of production estimation (both variants) can be applied to animal populations possessing more or less easily distinguishable age-groups (for example, crustaceans); it will prove difficult to use for animals such as rotifers with no markedly distinct developmental stages.

All the methods for calculating production examined here have been developed only in very recent times, as has already been pointed out. They have involved making many assumptions so that it is very important that future investigators should aim to verify and perfect them and to increase their precision. It is absolutely essential that further studies should be made on such highly important indices as the rate of development of a given species in relation to temperature and food conditions, in order to establish the growth characteristics and the normal patterns of food consumption and assimilation in animals from different taxonomic groups.

F

Chapter 6

THE PLACE OF THE PRODUCTION OF SPECIES
POPULATIONS IN THE COMMUNITY

6.1 General Remarks

For many invertebrates playing an important rôle in the processes of production, the most common factor causing their elimination is that they are fed upon by animals from the next trophic level. This circumstance provides the possibility of assessing that fraction of the prey-population production which goes to satisfying the food requirements of the next trophic link. Any quantitative estimate of mortality by predation gives us some idea of where and how effectively is the production of the various species-populations utilized within the community. It also enables us to define more precisely the functional rôle of a particular population in the ecosystem.

Investigations on the food requirements of aquatic invertebrates have been carried out for some time, chiefly in fisheries work and in experimental studies on the food habits of predatory invertebrates. However, in only a few cases can the results from such research be used to estimate the potential productivity of the consumed organism. Obviously, the more fully the consumed part of a given invertebrate prey species-population is known and the more precisely the ration of the predator species-population is determined, the closer will our approximate estimation of the production of the consumed prey species be to realistic values.

Although there were several other research workers who earlier compared the food rations of fish with the biomass of their food species, it was the New Zealand investigator, Allen (1945, 1951), who was one of the first to draw attention to the possibility of estimating the productivity of prey species from the quantities consumed by fish. Allen investigated a population of trout in the Horokiwi River. In addition to establishing the numbers of fish present in a reach of the river, their growth rates and the age structure of the population, he also studied both the composition of the food and the quantities eaten by these fish. From this information he was able to discover the type and number of invertebrates needed to satisfy the food requirements of the fish population. From a comparison of these values with calculated biomass levels of invertebrates present in the reach of the river being studied, he concluded that the production of the food species exceeded their biomass by 40–150 times.

Gerking (1954, 1962, 1964) used a similar approach in studying the productivity and efficiency of food utilization of a population of big-eared perch (*Lepomis macrochirus*).

Mann (1964) examined the energy flow through the animal inhabitants (fish and invertebrates) in a reach of the River Thames and concluded that, during one year, the fish consumed invertebrates in amounts which were 18 times their mean annual biomass.

In the Soviet Union similar approaches to the problem of determining what fraction of invertebrate production is used by fish can be found in the work of Vorob'ev (1949), Shorygin (1952), and others. These authors have used data on the amount of a food species consumed by fish to estimate approximate P/B coefficients for marine benthic invertebrates. Lyakhnovich (1958, 1961), Maksimova (1961) and Winberg and Lyakhnovich (1965) have all examined the production of invertebrates which supply the food requirements of fish in carp ponds.

More and more research workers are being attracted by the perspective offered by this approach of a quantitative evaluation of the significance of various species-populations in ecosystems.

6.2 Some Examples of Estimates of Benthic Production made from Measurement of Food Consumed by Fish

One of the first Soviet workers to make tentative production estimates of the benthos of the Azov Sea from the measured food consumption of benthophagous fish was Vorob'ev (1949). In practice, what he actually determined was the fraction of benthic production which served as food for these bottom-feeding fish. Over a period of several years, Vorob'ev carried out extensive investigations on the composition, distribution, numbers and biomass of organisms living on the bottom and, from this, established their general stock levels, seasonal changes and the relative weights within the general biomass of the predominant forms and groups. Simultaneously, analysis of the gut contents of the benthophagous fish established the percentage frequency of different invertebrates in their food. Making the assumption that the benthophagous fish stocks (ichthyomass) in the Azov Sea were about $600\,000T$ and that the fish could eat during a year a quantity of benthic organisms that was 23 times their standing crop, the annual food requirements of the total fish stocks could be assessed. This was converted into units of surface area (per square metre) and arranged into the various taxonomic groups contributing to the food supply according to their relative weight frequency in the gut contents of the fish.

Despite the continual fish cropping of the benthos, these organisms increased their standing crop throughout the summer. The difference between the autumnal and spring levels of invertebrate biomass was named the "residual production" by Vorob'ev, which, summed with the quantity eaten by the fish, he considered to give the total production. The ratio of this total production to the spring biomass provided P/B coefficients for the whole benthos as well as for the various predominant benthic groups (Table 6.1). This term "'residual production'', denoting the biomass increase during the period of the investigation, is considered to be unnecessary, although, of

TABLE 6.1

Approximate estimates of mean benthic production (Vorob'ev, 1949)

Taxonomic group	Mean biomass		Summer biomass increment (g m⁻²)	Consumption by fish (g m⁻²)	Production without mortality (g m⁻²)	P/B
	Spring (g m⁻²)	Summer (g m⁻²)				
Bivalve molluscs	124	299	175	197	372	3
Gastropod molluscs	3	4	1	26	27	9
Crustaceans	5	10	5	60	65	13
Balanus	6	27	21	21	42	7
Worms	7	8	1	69	70	10
TOTAL	145	348	203	373	576	42

course, the biomass increase itself, along with any other changes in biomass, must be included in all production calculations.

Vorob'ev managed to determine the production of certain Azov Sea molluscs by another method, a more accurate one in his opinion. This can be illustrated by the example of *Cardium*. The author noticed that the percentage frequency of the O-group *Cardium* decreased as one moved from regions with low fish exploitation to others where large fish shoals remained for long periods of time. By assuming both that the settlement of *Cardium* larvae and that the natural mortality of the molluscs was similar in all parts of the Sea, and also that few young *Cardium* were consumed in the northern parts of the Sea where there were few fish, Vorob'ev was able to calculate for the entire Sea the average mortality of O-group molluscs due to consumption by fish for the period from spring to autumn (Table 6.2).

The mean number of O-group *Cardium* estimated from dredge samples was 486 individuals per square metre in the autumn which, from Table 6.2,

TABLE 6.2

Mortality in young molluscs in relation to fish abundance (Vorob'ev, 1949)

Region	Abundance of O-group Cardium		Abundance of fish	Percentage loss of young molluscs, compared with the loss in regions without fish consumption
	Frequency (%)	Individuals per square metre		
I	12·42	177	many	77·70
II	17·86	255	fair numbers	67·88
III	25·79	368	less	53·65
IV	38·97	556	few	29·97
V	55·69	794	none	0·00
			MEAN	57·3

represents 42·7 per cent (i.e. $100 - 57·3$ per cent). The number consumed by fish (57·3 per cent) was therefore 652 individuals per square metre. Knowing the mean individual weight of the O-group molluscs, the author calculated that the fish consumed 103·7 g m^{-2}. The number of older *Cardium* consumed by fish was assumed to be in the same proportion to the O-group animals as was found in the fish stomachs. From all these calculations, Vorob'ev was able to determine that the P/B coefficient for *Cardium* was 4·25, and similarly 3·22 for *Mytilaster lineatus* and 2·05 for *Syndesmus ovata*. The mean P/B coefficient for all three species was 3·17, which is very close to the value obtained for the bivalves in Table 6.1, calculated by the other method.

Shorygin (1952) studied the food habits and trophic interrelations between the fish of the Caspian Sea and, like Vorob'ev, attempted to estimate the levels of benthic production providing food for the fish. The ratio between the consumed quantities and their biomass (standing crop) of the various benthic groups involved was considered, correctly, by Shorygin to provide a minimal estimate of their P/B coefficients (Table 6.3).

The P/B coefficient for the gastropod molluscs seems to be an overestimate and chironomid larvae are apparently insufficiently exploited by the fish of the northern Caspian. Shorygin does not describe how he calculated either the total fish stocks or their annual food requirements (rations) so that it is difficult to judge what were the possible sources of error. However, it is interesting to note that the P/B coefficient arrived at by Shorygin for the whole zoobenthos of the northern Caspian was similar to that calculated by Vorob'ev for the benthos of the Azov Sea.

Apart from the various assumptions made during the course of the

TABLE 6.3

Minimal estimates of P/B coefficients for benthic organisms (Shorygin, 1952)

Organisms	P/B	Organisms	P/B
Crustaceans	17·0	*Mytilaster*	9·0
Didacna	0·5	Gastropods	20·0
Adacna	4·6	All molluscs	3·0
Monodacna	2·4	Worms	3·2
Cardium	3·5	Chironomids	2·5
Dreissena	5·7	Total zoobenthos	4·0

calculations, the weakest point lies in the estimate of the fish biomass; it is very difficult to make even the crudest approximation of this value in natural water bodies. For this reason it is important that the values of production obtained by the method just described be compared with values calculated by some of the procedures outlined in the previous chapters.

6.3 Estimation of Invertebrate Production in Lakes from their Consumption as Food

When attempts were made to calculate the total quantities of food consumed by fish populations of lakes over a period of time, the difficulties encountered in estimating the absolute fish numbers and weight were similar to those for fish populations in the sea. However, sometimes these difficulties may be overcome.

Dadikyan (1955a, b) estimated the quantities of invertebrates consumed by the fish of Lake Sevan by applying a method similar to that of Vorob'ev (1949) and Shorygin (1952). After a thorough study of the food eaten by trout of various ages (starting with two-year-olds) lasting a few years (1948–1951), Dadikyan established the ratios between the various components of the trout diet. He assessed the quantity of food consumed by the trout from the daily rhythm in feeding from which he obtained some idea about the number of times per day the fish filled and emptied its gut. With this information, together with a number of assumptions about the seasonal changes in the feeding intensity of trout, he estimated that trout of different breeds consumed in a year a quantity of food which was 6·38 to 7·80 times greater than their own mean weight.

The total number of fish in the lake was calculated by Dadikyan from the records of fishery catches, available for many years and analysed into age classes. From these calculations, it was estimated that trout consumed $739·2T$ per year in Lake Sevan. Other fish species were present in the lake but in insignificant numbers and their food consumption was not very great; how-ever, the amount of food they consumed was calculated in the same way as for trout. The total quantity of invertebrates in the lake consumed by fish was then separated into groups according to their relative weight frequency in the fish gut contents. The values finally obtained were then compared with the levels of biomass or production of Lake Sevan invertebrates found by other workers.

Dadikyan demonstrated that fish more than two years old used only 2 per cent of the zooplankton production (determined by Meshkova, 1952) but that the production of *Gammarus*, determined by Markosyan (1948), was being completely consumed. The consumption by fish of other groups of invertebrates was not related to their production but to their mean annual

TABLE 6.4

The biomass of various invertebrates in Lake Sevan and the quantities consumed by fish (Dadikyan, 1955), in tonnes

Indices	Oligochaeta	Chironomid larvae	Leeches	Trichoptera	Molluscs
Annual biomass	2998	976	383	56·1	16·4
Consumed by fish	—	62	45	120·7	15·9
P/B	—	0·06	0·12	2·2	0·96

biomass, determined by other research workers at the Hydrobiological Station on Lake Sevan (Table 6.4).

The first three groups of invertebrates in Table 6.4 were not eaten to any considerable extent in Lake Sevan by the fish which were two years old or more, so that one cannot estimate their production using these indices. But that part of the production of Trichoptera and molluscs which was consumed by fish represents a value of 2·2 or 0·96 respectively of their mean annual biomass. This demonstrates clearly the extent of the invertebrate contribution to fish production in the lake.

It was possible to determine with far greater accuracy the chironomid production consumed by fish in the hatching and rearing lakes of the Tep-lovski Fishery Factory (Levanidova, 1959; Levanidov and Levanidova, 1962). In these waters it was quite possible to measure not only the total fish standing crop but also the fish production from growth increments over a period of time. The amount of food eaten by fish per unit of fish growth (the food coefficient) was determined experimentally. Finally, calculations were made on the production of chironomid larvae consumed by the chum salmon (*Oncorynchus keta*) and other fish fry in the Teplovka River and Teploye

Lake. These authors then added this part of the chironomid production consumed by fish to the biomass of chironomid larvae, prior to adult emergence and considered this to represent the total chironomid production, apart from losses due to natural mortality.

The chironomid larvae which were predominant in the water bodies being investigated belonged to the Chironomidae, the Diamesinae and the Orthocladiinae. Taking all the chironomids together, the ratio of calculated production to their spring biomass was 4·3.

Working in small North American lakes inhabited mostly by the big-eared perch (*Lepomis macrochirus*), Gerking determined the quantities of invertebrates consumed by fish, using special methods (Gerking, 1954, 1962, 1964). Together with detailed studies on the composition of the food eaten by the fish, he paid particular attention to determining the fish biomass of the lake and their production and losses due to fishing and natural mortality. The total number of fish in the lake was estimated by releasing tagged fish and recording the subsequent return; the percentage of tagged fish returned in the commercial catch enabled the total fish stock to be calculated. The growth rate of the fish was obtained by back-calculating from scale readings. Some estimations were also made of recruitment rates and rates of mortality due to fishing activity and to natural causes.

Gerking determined the food ration of big-eared perch experimentally from the degree of protein utilization, using the Kjeldahl technique to measure the nitrogen content of the food, faeces and the fish body itself at the beginning and end of each experiment. The results were converted to wet weight of food species, assuming a mean nitrogen content of 9·59 per cent.

The completed calculations revealed that in Lake Wayland during July the fish consumed a quantity of invertebrate species which was six times greater than their biomass. The chironomid larval production consumed during July by the fish population was 3·6 times the insect standing crop. In 1962 Gerking critically re-examined the calculations of Allen (1951) and showed that his levels of possible invertebrate production in the Horokiwi River were considerably overestimated.

During an investigation on how much of the invertebrate production was used to satisfy the food requirements of the fish living in a reach of the River Thames, Mann (1964) determined the fish ration from the balance equation of Winberg (1956), which derives the food ration of fish from their growth and metabolic rates. Mann demonstrated that the smallest possible P/B coefficient for the benthic invertebrates (with the exception of the large bivalves) could not be less than 18.

It is well known that the production of aquatic invertebrates is consumed not only by fish but also by predatory invertebrates that belong to the same trophic level as the so-called non-predatory fish which themselves also feed on invertebrates. And yet the invertebrate predators occupy a different position in the food chain since they are being preyed upon by the non-predatory fish.

As to how much of the invertebrate production goes to satisfy the food requirements of invertebrate predators, we have as yet very little information for realistic conditions. A number of authors have shown that, under laboratory conditions, many invertebrate predators are able to exterminate their prey in enormous quantities, although most predators do not usually wholly consume their prey. Should the potential capacity of invertebrate predators in this respect be given free reign under natural conditions, there would be little food left over for fish, apart from the predators themselves.

An attempt has been made by Shushkina (1964a) to evaluate the rôle of planktonic invertebrate predators. The eutrophic Lake Batorin (Belorussian SSR) has a summer level of zooplankton biomass of 3·3–7·3 g m^{-3}. Present in the zooplankton were predatory Cyclopidae and *Leptodora*. Shushkina's results show that the predatory cyclopoids formed about 30 per cent of the zooplankton biomass, with *Mesocyclops leukarti* (Claus) being the most predominant of the four other Cyclopidae species present.

During the summer of 1962, Shushkina experimentally determined the food rations of predatory cyclopoids in Lake Batorin. Concentrated lake zooplankton provided the food supply and was isolated in experimental bottles to which were added a known number of adult cyclopoids and their copepodids. The bottles were suspended in the lake for two to three days, whereupon, after fixation, the loss in zooplankton during the period of the experiment was calculated and related to the numbers of cyclopoids introduced. Shushkina showed that rotifers did not form an important fraction of cyclopoid diet; the main food of predatory copepods were the cladocera. Under conditions of concentrated food and dense aggregations of predators, resembling those of a eutrophic lake, the daily food consumption of the predatory cyclopoids was 26–39 per cent (average 35 per cent) of their wet weight.

A comparison of these results with the actual values of zooplankton biomass in the lake led Shushkina to conclude that copepodid and adult cyclopoids daily consumed 12–17 per cent of the biomass of the food species or 8–12 per cent of the total zooplankton biomass. Having calculated the consumption possible for *Leptodora*, which could attain densities of 3000 individuals per cubic metre, Shushkina's counts revealed that the crustacean planktonic predators all together consumed 13–22 per cent of the zooplankton biomass of Lake Batorin. Thus, if only to satisfy the food requirements of the planktonic predators, the daily production of the zooplankton must average about 18 per cent of its biomass and, from this, its P/B coefficient for a summer month had to be 5·4. If the fish ate roughly the same quantities of zooplankton (the commercial catch of fish in that lake was about 25 kg ha^{-1}), then the zooplankton monthly P/B coefficient must be between 10 and 12 during the summer season.

Following on from these results on the production of zooplanktonic predators and their possible rations calculated from Winberg's formula (1956), Shushkina (1966) indicates that the food requirements of planktonic predators can exceed the production of the filtering crustacean plankton in lakes of different type. A comparison of the potential production of herbivorous zooplankton (protozoa, rotifers, cladocerans, diaptomids) with the food consumption of their predators showed that the predators could crop 50 per cent of the production of the non-predatory zooplankton in lakes approaching oligotrophy, 65 per cent in mesotrophic lakes and about 40 per cent in eutrophic ones. The fraction of the zooplankton production which could be eaten by fish or other non-planktonic animals was estimated from the formula

$$P = P_0 - R_e + P_e \tag{6.1}$$

where P_0 is the summed productions of all the non-predatory zooplankton species-populations (belonging to first trophic level), R_e is the food consumed by the predatory zooplankton and P_e is the production of these predators (second trophic level).

Shushkina named the values obtained in this way the "real" production, since a straight summing of productions from two trophic levels would overestimate the food available to the fish feeding on plankton. But then any simple addition of the productions from different trophic levels is absurd and generally impossible, so that there is no need to single out the zooplankton or to introduce the concept of its "real" production—the more so because many non-planktonic invertebrates are consumers of the zooplankton in addition to the fish and zooplanktonic predators.

This approach of estimating that part of a population's production which goes to satisfy the food requirements of the succeeding trophic level (the predators) is yet in its infancy. As yet we have insufficient information about feeding types, about size of ration required by various predators among the invertebrates or about the invertebrates food requirements of fish. Nevertheless, even these few preliminary attempts show promise that proper investigation will enable us to comprehend the rôle a given species population plays within a community or within the ecosystem of the water body as a whole.

6.4 Production Estimation in Fish Ponds from Values of Consumption of Invertebrate Food Species

Many of the difficulties associated with attempts to determine the total number of fish in a habitat which are consuming the invertebrates can be solved easily in fish ponds, which are usually stocked with known numbers of fish of a known age and identity; moreover, growth rate can be measured at regular intervals. These conditions enable reasonably reliable estimations to be made of invertebrate production from the quantities consumed by the domesticated fish. Many of these drainable ponds can act as natural models for aquatic habitats and they are more suitable than any other type of water for productivity research. This is worth bearing in mind when productivity studies in fresh waters are being organized.

Repeated attempts have been made to estimate the proportion of invertebrate production feeding the fish in fish ponds and other small water bodies by isolating control areas of either benthos (Lyakhnovich, 1958; Lellak, 1957, 1961; Assman, 1960) or of plankton (Yaroshenko and Naberezhnyi, 1955). These experiments, testing the effects of excluding fish consumption from certain parts of the pond on the bottom or mid-water, produced unexpected results because the cropping by fish of a certain proportion of an invertebrate population stimulated its regenerative capacity (Winberg and Lyakhnovich, 1965; Lyakhnovich, 1967a, b). Even when intensively consumed, the number and biomass of invertebrates were not affected to any considerable extent and, in some cases, the invertebrate biomass was greater when being consumed by fish. In any case, it has proved impossible to make any estimations of invertebrate production which serves as invertebrate food in such isolated areas. This is clearly demonstrated by the results of investigations carried out simultaneously on ponds with different fish stocking densities (Gurzęda, 1960, 1965; Prosyanyi and Makina, 1960; Grygierek, 1962; Grygierek and Wolny, 1962; Grygierek et al., 1961) or ponds stocked with fish and ponds without fish (Hrbáček et al., 1961; Hrbáček, 1962; Hrbáček and Novotná-Dvořáková, 1965; Winberg and Lyakhnovich, 1965; Lyakhnovich, 1967a).

Lyakhnovich (1958, 1961) estimated the invertebrate production of the zooplankton which is the natural food supply of the young carp fry in Belorussian fish-farming ponds. Using hatching ponds, he determined the composition, number and biomass of the zooplankton, the number, biomass and growth rate of carp fry and also identified the food eaten from gut analysis. The daily food intake of the fry was calculated from their growth and respiratory rates, according to the balance equation interrelating these values (Winberg, 1956). Lyakhnovich showed that, in these ponds, the carp fry used daily not more than 2·1–7·2 per cent of the mean zooplanktonic biomass while he was studying them. Nevertheless, the standing crop of the zooplankton declined sharply, suggesting that consumers other than fish, probably invertebrate predators, were cropping them; otherwise, no explanation could be found for such an abrupt decrease in the level of the zooplankton.

Maksimova (1961) used a similar approach to examine the fish ponds of the Ropsh Hatchery in the district of Leningrad. Here, according to her observations, the mean daily cropping of invertebrates by the O-group carp was calculated and is given as a percentage of the standing crop in Table 6.5.

The values listed in Table 6.5 were calculated from observations made during the same months—July, August and September—but in different years on six ponds. Maksimova's data appear to be too high, apparently because the invertebrates serving as food species were not estimated accurately enough.

TABLE 6.5

The mean daily feeding intensity of O-group carp given as percentage of the biomass of the invertebrate food species (Maksimova, 1961)

Organisms	The mean for all ponds (%)	The range for different ponds (%)
Daphnia	40·8	27·8–77·6
Eurycercus	46·5	17·3–71·1
Bosmina	35·2	—
Chydoridae	27·4	27·1–27·6
Other Cladocera	10·3	0·7–22·6
Cyclopidea	20·5	14·0–30·3
Diaptomus	12·5	0·0–48·4
Chironomus plumosus	8·4	2·7–19·4
Procladius	1·1	—
Chironomid larvae from weeds	18·4	13·8–24·4
Chironomid pupae	54·1	45·0–74·7
Ephemeropteran larvae	38·6	24·9–81·6
Other insect larvae	12·2	—
Other bethnic organisms	2·1	0·1–4·3

Another possible error might lie in an inaccurate estimation of food consumption which was based on growth rates and published values of metabolic rates for fish of a given age. A factor of 2 was used to correct for active metabolism, whereas for carp fry in ponds a factor of 1·5 applied to the normal rate of metabolism measured in a respirometer would have given a more reliable and realistic level.

Despite these possible errors and the incompleteness of Maksimova's data, it is at least revealed that, under conditions of intensive exploitation, the growth increment of zooplankton and benthos was very high in the ponds during the summer. It is particularly telling that the biomass of the zooplankton and zoobenthos in one pond without any carp fry (and so without fish consumption) did not, in fact, differ from those with fry.

The total quantity of food consumed by the fish population can also be calculated from the cumulated growth increments of the fish and the food coefficient which reflects the quantity of food consumed for a unit growth increment. This approach was used by Lyakhnovich (1961) to analyse his results from the fish-rearing and fattening ponds of Belorussia. Realizing that only a fraction and not the total invertebrate production could be evaluated by this method, he adopted a food coefficient of 4·5 for carp food of mixed zooplankton and zoobenthos and suggested that it could not be less. The quantity of invertebrates consumed was then obtained by multiplying the fish growth increment by this coefficient and then comparing the resulting value with the seasonal mean values of the zooplankton and zoobenthos being consumed by fish in the same ponds. It seems that 6·2–17·0 times the invertebrate mean biomass was eaten by the fish during the summer, and he considers that these values are the minimal possible P/B coefficients.

In more recent work on twenty ponds in Belorussia and five in Israel, Lyakhnovich calculated the quantities of food consumed at two food coefficient levels (F.C.), namely, 5 and 10 (Winberg and Lyakhnovich, 1965). When the food coefficient was 5, the mean P/B coefficient was 13·3, and when it was 10 the P/B coefficient was twice as high.

Here the assumption is made that in an intensely growing population of young fish (O-group and one-year-olds), the effective utilization of their natural food supply is likely to lie between these two values of the food coefficient. According to Maksimova's data, the total quantity of invertebrate food species was equal to 4·8–17·8 times the mean seasonal biomass of these invertebrates; the mean value for all six of these ponds was very close to that obtained by Lyakhnovich using a food coefficient of 5.

It is a peculiar feature of these methods of calculating the consumed invertebrate production of fish ponds that the calculated values of elimination cannot be related to the initial spring biomass because the ponds in which the experiments were carried out are regularly drained and the standing crop is therefore newly formed every time the pond is refilled. During the growing season, the biomass of both zooplankton and zoobenthos show sharp fluctuations and it is advisable to compare losses due to mortality with the mean biomass during the period of observation.

These fish pond investigations provide quite realistic data on how much production is channelled into satisfying the nutritional requirements of fish as well as a well-defined conception about how effectively growing fish make use of invertebrate production.

We can summarize the account given here about the ways of evaluating the productivity of various organisms, groups or entire communities of aquatic invertebrates from the intensity of their predators' feeding. During the period of work, whether a whole year or a growing season, the identity and quantity of invertebrates being consumed by a given population of predators (such as fish) needs to be determined throughout the predators' habitat. Simultaneously, some estimate must be made of the absolute number and age-size structure of the predator population, as well as their mode of feeding. From analysis of gut contents, some idea can be obtained about which species are eaten as well as their size or weight. From this data it is possible to learn something about the specific weight of each food species in the diet of the predator population.

It is not possible, by direct observation, to determine quantitatively the food consumption (ration) of a consumer population under field conditions of a water body, so that the ration must be estimated in one of several ways. It has often been feasible to use information about the daily feeding rhythm of a fish population for this purpose or indices on the time taken to fill the gut (Novikova, 1949, 1951; Dadikyan, 1955a). However, it may not always be possible to detect a daily feeding rhythm, and some species do not exhibit one. In such cases one must apply results on the sizes of food ration determined experimentally, either by direct measurement of food eaten or by the nitrogen balance method, to field populations in order to estimate the population food consumption.

Once the population growth increment or production of a consumer is known, as well as its food coefficient (the quantity of food expended per unit increment of growth), the quantity of food consumed by the population as a whole can be obtained by multiplying these two values together. Another more reliable and relatively simple way of calculating the population food ration is by means of Winberg's balance equation (Winberg, 1956), which states that "energy in the food ration = 1.25 (metabolic energy + growth increment energy)"; this requires information about the metabolic intensity and growth rates of the species. It is a method being employed ever more frequently for fish and aquatic invertebrates in hydrobiological and fishery investigations (Lyakhnovich, 1958; Maksimova, 1961; Ivlev, 1962; Mann, 1964; Shushkina, 1966). In studies on fish rations it is necessary to take into account that the metabolic rate of the less active fish will be only 25–50 per cent higher than the laboratory measurement indicates, whereas in the more active fish it will be 50–100 per cent higher.

Having obtained this data on the ration of the species being studied, together with quantitative information about its food composition, it is possible to calculate the total number of individuals and biomass of the food species population which has actually been consumed. A comparison of the values so obtained with quantitative field data for the same invertebrate species gives what proportion of either the initial or mean biomass of the food species has gone to satisfy the food needs of a consumer population during a given period of time.

Where relatively long periods of time are involved, such as a year or a whole growing season, the amount of food consumed by a fish population is usually found to exceed by several times the mean biomass of the consumed species. This ratio between the consumer's food ration and the mean standing crop of its prey provides an indication of how many times the food supply is being renewed directly as a result of the predator consumption; it evaluates the fraction of prey production being removed for food by the predator population.

This cropping of the production of a species population for food by organisms from the subsequent trophic level plays a significant rôle in the process of production since it stimulates the reproduction of the consumed species. The stability of the ecological system is maintained by the very fact that the capacity of a species for growth and reproduction increases with more intense predator consumption. This increase is maintained up to the point where mortality exceeds the intrinsic reproductive potential of the prey species as realized under given field conditions. A confrontation of the total production of a species population, as estimated by any one of the methods described in earlier chapters, with that fraction of it supplying the food of organisms of the next trophic level enables us to judge in what direction, how intensely and how effectively is the production of the various species-populations in a community being utilized.

6.5 Models as a Means of Assessing the Productivity of Aquatic Communities

Our ultimate aim in hydrobiological research on inland waters is to obtain an optimal control of their natural resources. By control we mean active interference by man in the water body as a whole system, by fertilization, by farming fish, by fisheries. This aim can be expressed in terms of financial cost or other quantitative expressions of useful production taken by man from the water body (for example, fish production). In natural water bodies, this control aims at obtaining the maximal values of total production for longer periods of time (in any event, a period of time longer than the life-span of the longest living inhabitant of the water body).

Interpreted in this way, the problem being examined comes down to an optimal management of an extremely complex system. An effective solution is possible only after constructing a model of the system to be managed—a model good enough to simulate the functioning of an aquatic ecosystem.

Until now only isolated attempts at modelling of this kind have been made (Karpov, Kroghius, Krokhin and Menshutkin, 1966; Winberg and Anisimov,

1966). All these models were executed on electronic computers of medium or higher capacity. The models were either systems of differential equations, solved by numerical methods, or direct finite-diverse systems with discrete time. It is also possible to build an ecosystem model as a composite of a number of finite automata.

As experience in this field is limited, it is difficult to generalize or to recommend which type of model to select. We can only point to the potentialities of electronic computers without thereby minimizing the possibilities of other methods (for example, analogue modelling). An expedient approach would be to develop a set of standard operations for electronic computers, which would define biological processes such as "reproduction", "mortality by predators" and so on; such an attempt has already been made but is applicable only to fish populations (Menshutkin, 1966a, b).

The construction of models to simulate ecosystems has to be based upon a harmonious combination between energetics and population dynamics. Modelling must develop towards abstract constructions which express the normal patterns of ecosystem dynamics, as well as describing specific ecosystems such as the ecosystem model of Lake Dal'neye in Kamchatka (Karpov *et al.*, 1966). Having designed a model for such a concrete aquatic ecosystem, it would be sensible to use it to plan further hydrobiological and ichthyological research there.

It should be stressed that models of an ecosystem can only be constructed if reliable information is available on various processes such as growth, reproduction, mortality, food habits and various other factors for all the species-populations composing the system. The model should be an abstraction as well as a synthesis of all the known information of the biological phenomena occurring in the body of water. Such an approach makes great demands on the original data on all points of energy transfer, it stimulates further research in the directions necessary for modelling and confronts investigators with the need to have comparable original data. Apart from the main aim of management of an ecosystem, the construction of ecosystem models provides an effective way of verifying whether or not its component experimental facts and observations are mutually contradictory.

Thus, the building of models can be a significant means for studying quantitatively the normal patterns of biological productivity. An understanding of these general laws and patterns will lead to a fuller exploitation of biological resources and the productive potentialities of fresh waters for man.

REFERENCES

The titles of all Russian articles have been transliterated and translated. Where an English translation is available a reference number is given with the initials as shown below:

C–C   Cover to cover translation
JPRS   Joint Publications Research Service, NY
FBA   Freshwater Biological Association, UK
NIO   National Institute of Oceanography, UK
NLL   National Lending Library for Science and Technology, UK
FRBC   Fisheries Research Board of Canada
AEC   Atomic Energy Commission, USA

Ablyamitova-Vinogradova, Z. A. (1949). O khimicheskom sostave bespozvono-chnykh chernogo Morya i ego izmeneniyakh. *Trudy Karzdah. nauch. Sta. T.I. vyazems'koho*, **7**, 3–50. (Changes in the chemical composition of Black Sea invertebrates.)

Allen, K. R. (1945). The trout population of the Horokiwi River: an investigation of the fundamentals of propagation, growth and survival. Rep. Fish. N.Z., 33–40.

Allen, K. R. (1951). The Horokiwi Stream. *Bull. mar. Dep. N.Z. Fish.* no. 10.

Anan'ichev, A. V. (1961). Stravnitel'naya biokhimicheskaya kharakteristika nekotorykh presnovodnykh bespozvonochnykh i ryb. *Biokhimiya*, **26**, 18–30. (A comparative biochemical characterization of some freshwater fishes and invertebrates. Trans. C–C. Biochemistry N.Y.)

Anderson, R. O. and Hooper, F. F. (1956). Seasonal abundance and production of littoral bottom fauna in a Southern Michigan Lake. *Trans. Amer. microsc. Soc.* **75**, 259–270.

Assman, A. V. (1960). Izmenenie dostuposti lichinok Chironomidae pri vyedamii ribami. *Izv. Akad. Nauk SSR ser. biol.* **5**, 670–685. (Changes in the availability of larval Chironomidae from consumption by fish.)

Bekman, M. Yu. (1954). Biologiya *Gammarus lacustris* Sars pribaikal'skikh. *Trudy baikal'. limnol. Sta.* **14**, 263–311. (The biology of *Gammarus lacustris* Sars in water bodies of the Baikal region.)

Bekman, M. Yu. (1959). Nekotorye zakonomernosti raspredeleniya i produtsiro-vaniya massovykh vidov zoobentosa v Malom More. *Trudy baikal'. limnol. Inst.* **17**, 342–381. (Some regularities in the distribution and production of dominant zoobenthic species in the Baikal Maloye More.)

Bekman, M. Yu. (1962). Ekologiya i produktsiya *Micruropus possolskii* Sow i *Gmelinoides fasciatus* Stelb. *Trudy baikal'. limnol. Inst.* **2/22**, no. 1, 141–155. (Ecology and production of *Micruropus possolskii* Sow and *Gmelinoides fasciatus* Stelb.)

Bekman, M. Yu. and Menshutkin, V. V. (1964). Analiz processa produtsirovaniya y populatsii prosteishikh struktur. *Zh. obshch. Biol.* **25**, no. 3, 177–187. (An analysis of the production process in simple populations. Trans. JPRS 26245.)

145

Bělĕradeck, J. (1929). Sur la significance des coefficients de temperature. *Protoplasma*, **7**, 232–255.

Bělĕradeck, J. (1935). Temperature and living matter. Protoplasma-Monogr. **8**, Berlin.

Birge, E. A. and Juday, C. (1922). The plankton. 1. Its quantity and chemical composition. *Bull. Wis. geol. nat. Hist. Surv.* **64**, 1–222.

Birger, T. I. (1961). "Kormova Tsinnist' dlya rib Massovikh form Beskhrebetnikh Dnipra i Dniprovs'ko-Bugskogo Limanu." Kiev. ("The Nutritive Value of Dominant Invertebrate Species for Fish in the Dniepr and the Dniepr-Bug Estuary.")

Blegvad, N. (1928). Investigation of the bottom invertebrates in the Limfjord 1910–1927 with special reference to the plaice food. *Rep. Dan. biol. Stn.* **34**, 33–52.

Blunck, M. (1923). Die Entwicklung von *Dytiscus marginalis* L. von Ei bis zur Imago. *Z. wiss. Zool.* **71**, 171–391.

Bodenheimer, F. S. (1934). Uber die Temperaturabhangigkeit der Insekten. *Zool. Jb.* (*Zool. u. Physiol.*), **66**, 113–151.

Borutski, E. V. (1934). K voprosu o tekhnike kolichestvennogo ucheta donnoi fauny. Soobshch. III. *Trudy limnol. Sta. Kosino*, **18**, 109–132. (The problem of procedures for the quantitative estimation of bottom fauna.)

Borutski, E. V. (1939a). Dinamika biomassy *Chironomus plumosus* profundali Belogo ozera. *Trudy limnol. Sta. Kosino*, **22**, 156–195. (Dynamics of the biomass of *Chironomus plumosus* in the profundal water of Lake Beloye.)

Borutski, E. V. (1939b). Opredelenie produktivnosti bentosa ozer putem izucheniya dinamiki biomassy. Dokt. Diss. (Determination of lake benthos productivity from knowledge of the biomass dynamics.)

Boysen-Jensen, P. (1919). Valuation of the Limfjord. I. Studies on the fish food in the Limfjord 1909–1917. *Rep. Dan. biol. Stn.* **26**, 1–44.

Brandt, K. and Raben, E. (1919). Zur Kentmiss der chemischen Zusammensetzung des Planktons und einiger Bodenorganismen. *Wiss. Meeresunters. Abt. Kiel.* **19**, 175–210.

Bregman, Yu. E. (1968). Rost i produktsiya kolovratki *Asplanchna priodonta* v evtrofnom ozere Drivyat'. *V nast. sb* 184–193. (Growth and production of the rotifer *Asplanchna priodonta* in the eutrophic Lake Drivyaty. *In* "The Methods for the Estimation of Production of Aquatic Animals." Handbook and papers pt. II, paper 2. ed. G. G. Winberg. Minsk. Trans. FRBC 1172.)

Briskina, M. M. (1950). Materialy po biologii razvitiya i razmnozheniya nekotorykh morskikh i solonovatovodnykh amfipod. *Trudy karadah. nauch. Sta. T.I. Vyazems'koho* **10**, 46–59. (Contributions relating to the developmental biology and reproduction of some marine and subsaline Amphipoda.)

Brody, S. (1945). "Bioenergetics and Growth." Reinhold Publishing Co., New York.

Brotskaya, V. A. and Zenkevich, L. A. (1936). Biologicheskaya produktivnost' morskikh vodoemov. *Zool. Zh.* **15** no. 1, 13–25. (Biological productivity of sea waters.)

Brown, L. A. (1929). The natural history of cladocerans in relation to temperature. II. Temperature coefficients for development. *Am. Nat.* **63**, 346–352.

Burshtein, A. I. (1963). "Metody Issledovaniya Pishchevykh Produktov." Kiev. ("Methods of Studying Food Products.")

Chmyr, V. D. (1967). Radiouglerodnyi metod opredeleniya produktsii zooplanktona v estestvennoi populyatsii. *Dokl. Akad. Nauk SSSR*, **173**, 201–203. (The radio carbon method of determining the production of zooplankton in a natural population. Trans. C-C. and FBA trans. selected sections.)

Clarke, G. L. (1946). Dynamics of production in a marine area. *Ecol. Monogr.* **16**, 321–335.

Clarke, G. L., Edmondson, W. T. and Ricker, W. E. (1946). Mathematical formulation of biological productivity. *Ecol. Monogr.* **16**, 336–338.

Coker, R. E. (1933). Influence of temperature on size of freshwater Copepoda (Cyclops). *Int. Revue ges. Hydrobiol. Hydrogr.* **29**, 406–436.

Comita, G. and Anderson, G. (1959). The seasonal development of a population of *Diaptomus ashlandi* Marsh and related phytoplankton cycles in Lake Washington. *Limnol. Oceanogr.* **4**, 37–52.

Comita, G. and Schindler, D. W. (1963). Calorific values of microcrustacea. *Science N.Y.* **140**, 1391–1396.

Conover, R. (1962). Metabolism and growth in *Calanus hyperboreus* in relation to its life cycle. *Rapp. P.-v. Réun. Cons. perm. int. Explor. Mer*, **153**, 190–197.

Cooper, N. E. (1965). Dynamics and production of a natural population of a freshwater amphipod, *Hyalella azteca*. *Ecol. Monogr.* **35**, 377–394.

Crisp, D. T. (1962). Estimates of the annual production of *Corixa germani* (Fieb.) in an upland reservoir. *Arch. Hydrobiol.* **58**, 210–223.

Dadikyan, M. G. (1955a). Pitanie Sevenskikh forelei. *Trudy sevan. gidrobiol. Sta.* **14**, 5–76. (The nutrition of trout in Lake Sevan.)

Dadikyan, M. G. (1955b). Opyt opredeleniya velichiny godovogo potrebleniya korma Sevenskimi forelyami i vyedaemosti otdel'nykh predstavitelei bentosa. *Izv. Akad. Nauk armyan. SSR* **8**, 3–18. (Experiment to determine the annual food consumption of Sevan trout and the edibility of different benthic groups.)

Deevey, G. B. (1960). Relative effects of temperature and food on seasonal variations in length of marine copepods in some eastern American and western European waters. *Bull. Bingham. Oceanogr. Coll.* **17**, 54–86.

Drozdov, B. N. (1962). "Kalorimetr dlya Opredeleniya Teploty Sgoraniya Topliva." Moscow. ("Calorimeter for Determining the Combustion Heat of Fuel.")

Dunke, N. A. (1960). Nekotorye dannye o sostoyanii populyatsii *Bosmina longirostris* (O. F. Müller) v prudakh, udobrennykh raznym kolichestrom azotnofosfornykh solei. *Trudy belorussk. nauchno-issled. ryb. khoz.* **3**, 71–76. (Some data on the state of a *Bosmina longirostris* (O. F. Müller) population in ponds fertilized with varying quantities of nitrophosphoric salts.)

Edmondson, W. T. (1960). Reproductive rates of rotifers in natural populations. *Memorie Ist. ital. Idrobiol.* **12**, 21–77.

Edmondson, W. T. (1962). Food supply and reproduction of zooplankton in relation to phytoplankton population. *Rapp. P.-v. Réun. Cons. perm. int. Explor. Mer*, **153**, 137–141.

Edmondson, W. T. (1965). Reproductive rates of planktonic rotifers as related to food and temperature in nature. *Ecol. Monogr.* **35**, 61–111.

Edmondson, W. T., Comita, G. W. and Anderson, G. (1962). Reproductive rate of copepods in nature and its relation to phytoplankton population. *Ecology*, **43**, 625–634.

Eichhorn, R. (1957). Zur Populationsdynamik der Calaniden Copepoden in Titsee und Feldsee. *Arch. Hydrobiol. Suppl.* **24**, no. 2, 186–246.

Elster, H. I. (1954). Uber die Populationsdynamik von *Eudiaptomus gracilis* Sars und *Heterocope borealis* Fischer in Bodensee-Obersee. *Arch. Hydrobiol. Suppl.* **20**, 546–614. (Trans. FBA (New Series) 6.)

Elster, H. I. (1955). Zooplankton. Ein Beitrag zur Productionsbiologie des Zooplankton. *Verh. internat. Verein. Limnol.* **12**, 404–411.

Embody, G. (1912). A preliminary study of the distribution, food and reproductive capacity of some freshwater amphipods. *Int. Revue ges. Hydrobiol. Hydrogr. Suppl.* **3**, 1–33.

Faustov, V. S. and Zotov, A. I. (1965). Izmenenie teplot sgoraniya yaits ryb i amfibii vo vremya razvitiya. *Dokl. Akad. Nauch SSSR*, **162** no. 4, 965–968. (Changes in combustion heat of fish and amphibian eggs in the course of development. Trans. C-C.)

Finenko, Z. Z. (1965). "Soderzhanie Organicheskogo Veshchestva v Sestone Chernogo Azovskogo Morei." Kiev. ("Organic Matter in the Seston of the Black and Azov Seas.")

Fisher, L. R. (1962). The total lipid material in some species of marine zooplankton. *Rapp. P.-v. Réun. Cons. perm. int. Explor. Mer*, **153**, 129-136.

Friedemann, T. and Kendall, A. (1924). The determination of carbon and carbon dioxide. *J. biol. Chem.* **82**, 45–55.

Gaevskaya, N. S. (1938). O nekotorykh novykh metodakh v izuchenii pitaniya organizmov. 1. Opredelenie tochnogo vesa melikh vodnykh zhivotnykh v zhivom sostoyanii. *Zool. Zh.* **17**, no. 1, 165–174. (Some new methods of research into the nutrition of organisms. 1. Determining the accurate weight of small aquatic animals in their live state.)

Galkovskaya, G. A. (1963). Ob ispol'zovanii pishchi na rost i ob usloviyakh maksimal'nogo vykhoda produktsii kolovrathi *Brachionus calyciflorus* Pallas. *Zool. Zh.* **12**, no. 4, 506–512. (Food utilization for growth and conditions for maximum production of the rotifer *Brachionus calyciflorus* Pallas. Trans. FRBC 997.)

Galkovskaya, G. A. (1965). Planktonnye kolovratki i ikh rol' v produktivnosti vodoemov. Avtoref. Diss. Minsk. (Planktonic rotifers and their rôle in the productivity of water bodies. Author's report on her dissertation.)

Galkovskaya, G. A. and Lyakhnovich, V. P. (1966). Produktsiya prudovogo zooplanktona. Soobshch. 1. Produktsiya vetvistousykh rachkov *Daphnia pulex* (De Geer) i *Daphnia longispina* O. F. Müller v opytniykh prudakh (rybkhoz Izobelino). *Gidrobiol. Zh.* **2**, no. 4, 8-15. (The production of pond zooplankton. I. Production of the cladoceran crustaceans *Daphnia pulex* (De Geer) and *Daphnia longispina* O. F. Müller in experimental ponds (fish farm Isobelino).)

Gavrilov, S. I. and Arabina, I. P. (1967). Sootnoshenie vesa i lineinykh razmerov u predstavitelei presnovodnogo bentosa. *Gidrobiol. Zh.* **3**, no. 2, 71–73. (The correlation of weight and linear dimensions in representatives of the freshwater benthos.)

Geinrikh, A. K. (1956). O produktsii kopepod v Barentsevom More. *Dokl. Akad. Nauk SSSR*, III, no. 1, 199–201. (Copepod production in the Barents Sea. Trans. C-C.)

Gerking, S. D. (1954). The food turnover of a Bluegill population. *Ecology*, **35**, 490–498.

Gerking, S. D. (1962). Production and food utilization in a population of Bluegill Sunfish. *Ecol. Monogr.* **32**, 31–78.

Gerking, S. D. (1964). Timing and magnitude of the production of a Bluegill Sunfish population and its food supply. *Verh. Internat. Verein. Limnol.* **15**, 496–503.

Golley, F. B. (1961). Energy values of ecological materials. *Ecology*, **42**, 581–584.

Green, J. (1954). Size and reproduction in *Daphnia magna* (Crustacea: Cladocera). *Proc. zool. Soc. Lond.* **124**, 535–545.

Greze, B. S. (1948). Materialy po produktivnosti zooplanktona v Valdaiskom ozere. *Izv. vses. nauchno-issled. Inst. ozern. rechn. ryb. Khoz.* **26**, 2(11), 25–88. (Data on the zooplankton productivity in Valdai Lake.)

Greze, B. S. (1951). Produktsiya *Pontoporeia affinis* i metod ee opredeleniya. *Trudy vses. Gidrobiol. Obshch.* **3**, 33–43. (The production of *Pontoporeia affinis* and methods of determining it.)

Greze, B. S. (1963). Metod rascheta produktsiya planktonnykh kopepod. *Zool. Zh.* **17**, no. 9, 1329–1337. (A method of computing the production of planktonic Copepoda. Trans. JPRS 21801, NLL TT-63-41104.)

Greze, B. S. and Baldina, E. P. (1964). Dinamika populyatsii i godovaya produktsiya *Acartia clausi* Giesbr. i *Centropages kroyeri* Giesbr. v neriticheskoi zone Chernogo More. *Trudy sevastopol'. biol. Sta.* **17**, 249–261. (Population dynamics and annual production of *Acartia clausi* Giesbr. and *Centropages kroyeri* Giesbr. in the neritic zone of the Black Sea. Trans. FRBC 893.)

Grove, F. L., Jones, R. S. and Matthews, W. (1961). The loss of sodium and potassium during dry ashing of animal tissue. *Analyt. Biochem.* **2**, 221–230.

Grygierek, E. (1962). Wplyw zageszczenic narybku karpi na faune skorupiakow planktonowych. *Roczn. Nauk. roln.* 81-B-2, 189–210. (The effect of an increased density of carp fry on the planktonic crustacean fauna.)

Grygierek, E. and Wolny, P. (1962). Wplyw narybku karpia na jakość i liczebność wystepowania slimaków w malych stawach. *Roczn. Nauk. roln.* 81-B-2, 211–230. (The effect of an increase in carp fry on the quantity and quality of snails present in small ponds.)

Grygierek, E., Hillbricht, A. and Spodniewskaya, I. (1961). Izmeneniya planktonnogo biotsenoza, vyzvannye vliyaniem ryb, deistvuyushchikh kak khshchniki i kontroiryyushchikh sredu prudov. *Vop. Ekol.* **5**, 46–47. (Changes in the planktonic biocenosis caused by fishes acting as predators and controlling the environment of ponds.)

Gurzęda, A. (1960). Wplyw presji narybku karpi na dynamike liczebnośći Tendipedidae i Cladocera. *Ekol. pol.* **B, 6**, no. 3, 257–268. (The effect of carp stock pressure on numerical changes in Tendipedidae and Cladocera.)

Gurzęda, A. (1965). Density of carp populations and their artificial feeding and the utilization of food animals. *Ekol. pol.* **A, 13**, 73–99.

Hall, D. (1964). An experimental approach to the dynamics of a natural population of *Daphnia galeata mendotae*. *Ecology*, **45**, 94–112.

Hewitt, B. R. (1958). Spectrophotometric determination of total carbohydrate. *Nature*, **182**, 246–247.

Hrbáček, J. (1962). Species composition and the amount of the zooplankton in relation to fish stock. *Rozpr. cal. Akad. Ved.*, *Rada MPV*, **72**, no. 10, 1–116.

Hrbáček, J., Dvořáková, M., Kořínek, V. and Prochazkova, L. (1961). Demonstration of the effect of the fish stock on the species composition of zooplankton and the intensity of metabolism of the whole plankton association. *Verh. internat. Verein. Limnol.* **14**, 192–195.

Hrbáček, J. and Novotná-Dvořáková, M. (1965). Plankton of four backwaters related to their size and fish stock. *Rozpr. csl. Akad. Ved, Rada MPV*, **13**, no. 75, 3–65.

Hynes, H. B. N. (1961). The invertebrate fauna of a Welsh mountain stream. *Arch. Hydrobiol.* **57**, 344–388.

Ingle, L., Wood, F. and Banta, A. M. (1937). A study of longevity, growth, reproduction and heart rate in *Daphnia longispina*. *J. exp. Zool.* **76**, 325–352.

Ivlev, V. S. (1934). Eine Mikromethode zur Bestimmung des Kaloriengehalts von Nahrstoffen. *Biochem. Z.* **275**, 49–61.

Ivlev, V. S. (1938). O pretrashchenii energii pri roste bespozvonochnykh. Byull. *MOIP otd. Biol.* (*Bull. Soc. Nat. Moscou ser. Biol.*) **47**, no. 4, 267–277. (Energy transformation during the growth of invertebrates.)

Ivlev, V. S. (1939). Energeticheskii balans karpov. *Zool. Zh.* **18**, 308–318. (The energy balance of carp.)

Ivlev, V. S. (1945). Biologicheskaya produktivnost' vodoemov. *Usp. sovrem. Biol.* **19**, 98–120. (The biological productivity of waters. Trans. FRBC 394, W. E. Ricker (1966) *J. Fish. Res. Bd. Can.* **23**, 1727–1759)

Ivlev, V. S. (1962). Metod vychisleniya kolichestrav pishchi, potreblyaemoi rastushchei ryboi. *Biol. vnutr. vodoemov Pribaltiki Petrozavodsk.* 132–137. (A method of computing the food quantity consumed by a growing fish.)

Ivleva, I. V. (1953). Vliyanie pitaniya na intensivnost' razmnozheniya inkhitreid. *Trudy vses. nauchno-issled. Inst. morsk. ryb. Khoz. Okeanogr. Latviisk, otd.* **1**, 197–202. (The effect of nutrition on the reproductive intensity in Enchytraeidae.)

Janisch, E. (1932). The influence of temperature on the life history of insects. *Trans. R. ent. Soc. Lond.* **80**, 137–168.

Juday, C. (1940). The annual energy budget of an inland lake. *Ecology*, **21**, 438–451.

Juday, C. (1943). The utilization of aquatic food resources. *Science N.Y.* **97**, 456–459.

Kamshilov, M. M. (1951). Opredelenie vesa *Calanus finmarchicus* (Gunner) na osnovanii izmereniya dliny tela. *Dokl. Akad. Nauk SSSR*, **76**, no. 6, 945–949. (Determination of the weight of *Calanus finmarchicus* (Gunner) by the measurement of body length. Trans. NIO T/21, C-C.)

Kamshilov, M. M. (1958). Produktsiya *Calanus finmarchicus* (Gunner) v pribreznoi zone vostochnogo Murmana. *Trudy murmansk. biol. Sta.* **4**, 45–55. (The production of *Calanus finmarchicus* (Gunner) in the coastal zone of the eastern Murman.)

Karpov, V. G., Kroghius, F. V., Krokhin, E. M. and Menshutkin, V. V. (1966). A model of the ichthyocenosis of Lake Dalneye realised on an electronic computer. *Verh. Internat. Verein. Limnol.* **16**, 1095–1102.

Karzinkin, G. S. and Tarkovskaya, O. I. (1960). Opredelenie kaloriinosti v malykh naveskakh. *Trudy tomsk. gos. Univ.* **148**, 103–108. (Determining the caloricity of small weight samples.)

Keller, R. F. (1959). Color reactions for certain amino acids, amines and proteins. *Science, N.Y.* **129**, 1617–1618.

Khmeleva, N. N. (in press). Zatpaty energii na dykhanie, rost i kazmnozhenie u *Artemia salina* L. (The cost of maintenance, growth and reproduction in *Artemia salina* L.)

Kirpichenko, M. Ya. (1940). Donne tvarinne naselennya zaplavnikh vodoim r.

Dnipra Tsigan'ske i Pidbirne (v zv'yazku z vyyavlennem ikh tipu ta produktiv-nosti). *Trudy hydrobiol. Sta. Akad. Nauk Ukr. SSR*, **19**, 3–83. (Benthos populations in the flood basins of the Ts'gansko and Podborno regions of the River Dniepr—with regard to their species and productivity.)

Kizewetter, I. V. (1954). Kormovaya tsennost' planktona. *Izv. tikhookean. nauchno-issled. Inst. ryb. Khoz. Okeanogr.* **39**, 97–110. (The nutritional value of plankton.)

Kleiber, M. (1961). "The Fire of Life." New York.

Klekowski, R. and Shushkina, E. A. (1966). Energeteskii balans *Macrocyclops albidus* Jur. v period ego razvitiya. Ekologii vodnykh organizmov. Moscow, 125–136. (The energy balance of *Macrocyclops albidus* Jur. during the period of its development. *In* Ecology of aquatic organisms. Trans. FRBC 1031.)

Klekowski, R. and Shushkina, E. A. (1968). Pianie, rost i obmen *Macrocyclops albidus* za vremya ego metamorfoza pri raznykh kontsentratsiyakh korma. *Zh. obshch. Biol.* **29**, no. 2, 199–208. (Nutrition, growth and metabolism of *Macrocyclops albidus* during metamorphosis and at varying food concentrations. Trans. C-C.)

König, J. (1904). "Chemie der Menschilidren Nahrungs und Genussmitel." Berlin.

Konstantinov, A. S. (1951). O kolchestvennon uchete khironomid v pishche ryb. Metodika opredeleniya vozrasta lichinok. *Izv. vses. nauchno-issled. Inst. ryb. Khoz. Okeanogr.—Saratov. otd. kasp. fil.* **1**, 31–33. (The quantitative evaluation of Chironomidae in the food of fishes: Methods for determining the age of larvae. *see also* Trans. FRBC 1368.)

Konstantinov, A. S. (1958). Vliyanie temperatury na skorost' rosta i raznitiya lichinok khironomid. *Dokl. Akad. Nauk SSSR*, **120**, no. 6, 1362–1365. (The effect of temperature on growth and development rates of chironomid larvae. Trans. C-C.)

Konstantinov, A. S. (1962). Ves nekotorykh vodnykh bespozvonochnykh kak funktsiya ikh lineinykh razmerov. *Nauch. Dokl. Vyssh. Shkoly. biol. nauki.* **3**, 17–20. (The weight of some aquatic invertebrates as a function of their linear measurements.)

Konstantinova, N. S. (1961). O tempe rosta vetvistousykh rachkov i opredelenii okh produktsii. *Vopr. Ikhtiol.* **1**, 2, 19, 362–367. (The growth rate of cladoceran crustaceans and determination of their production. Trans. FRBC 410.)

Koshtoyants, Kh. S. (1950). "Osnovy Stravnitel'noi Fiziologii." T.1. Moscow. ("Foundations of Comparative Physiology.")

Kostylev, E. F. (1965). K voprosu opredeleniya summarnykh i udel'nykh kalorii-nostei biologicheskikh ob'ektov. *Gidrobiol. Zh.* **1**, no. 5, 61–65. (Problems in determining the total and relative calorific values of biological material.)

Kozhanchikov, I. V. (1946). K voprosu o zhiznennom termicheskom optimume. VIII. O labil'nosti protsessov razvitiya nasekomykh v otnoshenii termicheskikh vliyanii. *Zool. Zh.* **25**, no. 1, 27–34. (The question of the vital thermal optimum. VIII. The lability of development processes in insects in relation to thermal influences.)

Kozhov, M. M. (1950). "Zhivotn'i mir Ozera Baikal." Irkutsk. ("Baikal and its Life." Junk, Netherlands 1963.)

Krey, J. (1958). Chemical method of estimation of standing crop of phytoplankton. *Rapp. P.-v. Réun. Cons. perm. int. Explor. Mer*, **144**, 20–27.

Krogh, A. (1914). The quantitative relation between temperature and standard metabolism in animals. *Int. Z. phys.-chem. Biol.* **1**, 491–508.

Krogh, A. (1916). "The Respiratory Exchange of Animals and Man." Monographs on Biochemistry, Longmans, London.

Kryuchkova, N. M. and Kondratyuk, V. G. (1966). Zavisimost' fil'tratsionnogo pitaniya ot temperatury u nekotorykh predstavitelei otryada vetvistousykh rakoobraznykh. *Dokl. Akad. Nauk SSSR*, **10**, no. 2, 120–124. (The dependence of filter feeding on temperature in some forms of the Cladocera crustacean order. Trans. C-C.)

Kuznetsov, S. I. (1945). Biologicheskii metod otsenki bogatsgva vodoema biogennymi elementami. *Mikrobiologiya*, **14**, 248–253. (Biological methods of evaluating the abundance of biogene elements in water bodies.)

Kuznetsov, V. V. (1941). Dinamika *Microporella ciliara* v Barentsevom More. *Trudy zool. Inst. Leningr.* **7**, no. 2, 114–139. (Dynamics of *Microporella ciliata* in the Barents Sea.)

Kuznetsov, V. V. (1948a). Biologia i biologicheskii tsikl *Lacuna pallidula* v Barentsevom More. Sb. pamyati akademika Zernova. Moscow. pp. 72–93. (The biology and the biological cycle of *Lacuna pallidula* in the Barents Sea. Collection in memory of Academician Zernov.)

Kuznetsov, V. V. (1948b). Bio-ekologicheskaya kharakteristika massovykh vidov morskikh bespozvonochnykh, ch. 1. Biologicheskii tsikl *Lacuna vincta* (Montaga) na Vostochnom Murmane. *Trudy murmansk. biol. Sta.* **1**, 192–214. (A bio-ecological characterization of dominant pelagic invertebrate species. Pt. 1. The biological cycle of *Lacuna vincta* (Montaga) on the eastern Murman.)

Kuznetsov, V. V. (1948c). Bio-ekologicheskaya kharakteristika massovykh vidov morskikh bespozvonochnykh, ch. 2. Biologicheskii tsikl *Margarita helicina* (Phipp) Vostochnogo Murmane i Belnogo More. *Izv. Akad. Nauk SSSR, ser. biol.* **5**, 538–564. (A bio-ecological characterization of dominant pelagic invertebrate species. Pt. 2. The biological cycle of *Margarita helicina* (Phipp) of the eastern Murman and White Sea.)

Lebedeva, L. I. (1964). Opredelenie produktsii vetvistousykh rachkov v vodokhranilishchakh. Tezisy 10-i nauch. konf. po izuch. vnutr. vodoemov Pribaltiki, pp. 109–110. (Determining the production of Cladocera crustaceans in reservoirs. Xth Pribaltic conference on inland water studies.)

Lellak, J. (1957). Der Einfluss der Frestatigkeit des Fischbestandes auf die Bodenfauna der Fischteiche. *Z. Fisch.* **6** NF 8, 621–633.

Lellak, J. (1961). Zur Benthosproduktion und ihrer Dynamik in drei Bohmischen Teichen. *Verh. internat. Verein. Limnol.* **14**, 213–219.

Levanidov, V. Ya. and Levanidova, I. M. (1962). Nerestovo-vyrasnye vodoemy Teplovskogo rybovodnogo zavoda i ikh biologicheskaya produktivnost'. *Izv. tikhookean. nauchno-issled. Inst. ryb. Khoz. Okeanogr.* **48**, 3–66. (The hatching and rearing ponds of the Teplov fish farming plant and their biological productivity.)

Levanidova, I. M. (1959). Biologicheskaya produktivnost' nerestovo-vyrastnykh vodoemov tikhookeanskikh lososei na primere ozera Teploe. Trudy Soveshch. po probl. biol. vnutr. vodoemov, pp. 114–123. Akad. Nauk SSSR. Moscow and Leningrad. (The biological productivity of breeding and rearing ponds of the Lake Teploe Pacific salmon. VIth conference of the Problems of the Biology of Inland Waters.)

Lindeman, R. L. (1942). The trophic dynamic aspect of ecology. *Ecology*, **23**, 399–418.

Lohman, H. (1908). Untersuchungen zur Feststellung des vollstandigen Gehaltes des Meeres an Plankton. *Wiss Meeresunters. Abt. Kiel. N.F.* **10**, 131–370.

Lovegrove, T. (1962). The effect of various factors on dry weight values. *Rapp. P.-v. Réun. Cons. perm. int. Explor. Mer*, **153**, 86–91.

Lowry, O., Rosenbrough, N. and Pandau, R. (1951). Protein measurement with the Folin phenol reagent. *J. Biol. Chem.* **193**, 1–265.

Lundbeck, J. (1926). Die Bodentierwelt norddeutsch Seen. *Arch. Hydrobiol. Suppl.* **7**, 1–160.

Lusk, G. (1928). "The Science of Nutrition." Saunders, Philadelphia and London.

Lyakhnovich, V. P. (1958). Estestvennaya kormovaya baza ryb v prudovykh khozyaistvakh BSSR. 1. Kormovaya baza nerestovykh prudov i ee ispol'zovanie mal'kami. *Trudy belorussk. nauchno-issled. ryb. Khoz.* **2**, 150–199. (The natural supply of fish food in fish-farming ponds of the Belorussian SSR. 1. The food supply of the hatching ponds and its exploitation by fish fry.)

Lyakhnovich, V. P. (1961). Sootnoshenie kormovoi biomassy i ryboproduktivnosti v karpovykh. *Trudy vses. gidrobiol. Obshch.* **11**, 299–308. (The correlation between food biomass and fish productivity in carp ponds.)

Lyakhnovich, V. P. (1967a). Vliyanie dvuletnikh karpov na kolichestvennoe pazvitie zooplanktona i bentosa. Tr. 11-i nauch. konf. po izuch. vnutr. vodoemov Pribaltiki Petrozavodsk. (The effect of two-year-old carp on the quantitative development of the zooplankton and benthos. 11th Pribaltic Conference on Inland Waters.)

Lyakhnovich, V. P. (1967b). Zavisimost' kharaktera pitaniya dvuletnikh karpov ot sostava estestvennoi kormovoi bazy i vliyanie faktorov potrebleniya na dinamiku populyatsii zooplanktona i zoobentosa. Tr. 12-i nauch. konf. po izuch. vnutr. vodoemov Pribaltiki Vil'nyus. (The food habits of two-year carp related to the composition of the natural food supply and the effects of "grazing" on the population dynamics of the zooplankton and zoobenthos. 12th Pribaltic Conference on Inland Waters, Vilno.)

MacFadyen, E. (1965). "Ekologiya Zhivotn'kh. Tseli i Metod'." Moscow. ("Animal Ecology: Aims and Methods." 1957. Pitman, London. 2nd edition 1963.)

Maciolek, J. A. (1962). Limnological organic analyses by quantitative dichromate oxidation. *Res. Rep. U.S. Fish. Wildl. Serv.* **60**, 1–61.

Makhmudov, A. M. (1964). O khimicheskom sostave bentosa srednego i yuzhnogo kaspiya. *Zool. Zh.* **43**, no. 9, 1265–1275. (The chemical composition of the benthos in the central and southern parts of the Caspian Sea.)

Maksimova, L. P. (1961). Pitanie i stepen' ispol'zovaniya estestvennykh i iskusstvennykh kormov gibridami karpa s amurskim sazanom. *Izv. Gos. nauchnoissled. Inst. ozern. ryb. Khoz.* **51**, 65–95. (The feeding and degree of exploitation of natural and artificial foods by hybrids of carp—Amur carp.)

Malyarevskaya, A. Ya. and Birger, T. I. (1965). Khimicheskii sostav proizvoditelei, ikry i lichinok tarani i leshcha. Vliyanie kachestva proizvoditelei na potomstvo u ryb. Kiev, 5–34. (Chemical composition of spawners, eggs and larvae of taran (*Rutilus rutilus heckeli*) and bream—the effect of the spawners condition on the offspring in fish.)

Mann, K. H. (1964). The pattern of energy flow in the fish and invertebrate fauna of the River Thames. *Verh. Internat. Verein. Limnol.* **15**, 485–495.

Margalef, R. (1955). Temperatures, dimensions u evolucion. *Pros. Inst. Biol. Apl., Barcelona*, **19**, 13–94.

Markosyan, A. K. (1948). Biologiya gammarusov ozera Sevan. *Trudy sevan. gidrobiol. Sta.* **10**, 40–74. (The biology of Lake Sevan *Gammarus*.)

Marshall, S. (1949). Chitiruetsya po McLaren (1965). *Limnol. Oceanogr.* **10**, 525–538 (cited in McLaren (1965)—Marshall, S. (1949)). On the biology of the small copepods in Loch Striven. *J. mar. biol. Ass. U.K.* **28**, 45–122.)

Marshall, S. and Orr, A. P. (1952). On the biology of *Calanus finmarchicus*. VII. Factors affecting egg production. *J. mar. biol. Ass. U.K.* **30**, 527–547.

Mazepova, N. F. (1963). Biologiya pelagicheskogo rachka *Cyclops kolensis* Lill. v ozere Baikal. *Trudy baikal'. limnol. Inst.* **1**, no. 21, 49–134. (The biology of the pelagic crustacean *Cyclops kolensis* Kill. in Lake Baikal.)

McEvan, W. S. and Anderson, C. M. (1951). Miniature bomb calorimeter for the determination of the heat of combustion of samples of the order of 50 mg mass. *Rev. scient. Instrum.* **26**, 280–284.

McLaren, J. A. (1963). Effects of temperature on growth of zooplankton and the adaptive value of vertical migration. *J. Fish. Res. Bd. Can.* **20**, 685–727.

McLaren, J. A. (1965). Some relationships between temperature and egg size, body size, development rate and fecundity of the copepod *Pseudocalanus*. *Limnol. Oceanogr.* **10**, 528–538.

Mednikov, B. M. (1960). O produktsii kalanid severo-zapadnoi chasti Tikhogo-okeana. *Dokl. Akad. Nauk*, **134**, no. 5, 1208–1210. (Calanoid production in the north-western Pacific Ocean. Trans. C-C.)

Mednikov, B. M. (1962). O prodolzhitel'nosti metamorfoza beslonogikh rachkov i opredelenie produktsii vidov s rastyanutym periodom razmnozheniya. *Okeanologiya*, **2**, no. 5, 881–887. (The duration of metamorphosis in copepod crustaceans and the determination of production on species with a protracted period of reproduction. Trans. C-C.)

Mednikov, B. M. (1965). Vliyanie temperatury na razvitie poikilotermnykh zhivotnykh. 1. Pokazatel'nye gruppovye uraveniya razvitiya. *Zh. obshch. Biol.* **26**, no. 2, 190–200. (The effect of temperature on the development of poikilotherm animals. 1. Exponential group equations of development.)

Mendel, B., Kemp, A. and Mayerr, D. H. (1954). A colorimetric micro-method for the determination of glucose. *Biochem. J.* **56**, 639–646.

Menshutkin, V. V. (1966a). Ob optimal'noi strategii rybolovstva. *Vopr. Ikhtiol.* **6**, no. 38, 3–13. (The optimal strategy of angling fisheries.)

Menshutkin, V. V. (1966b). Optimal'noi upravlenie populyatsei promyslovoi ryby. *Vopr. Ikhtiol.* **6**, no. 41, 735–738. (The optimal management of a population of commercial fish.)

Meshkova, T. M. (1952). Zooplankton ozera Sevan (biologiya i produktivnost'). *Trudy sevan. gidrobiol. Sta.* **13**, 5–170. (The zooplankton of Lake Sevan (biology and productivity).)

Mitchell, W. H. (1929). The division rate of *Paramecium* in relation to temperature. *J. exp. Zool.* **54**, 383–410.

Mordukhai-Boltovskoi, F. D. (1949). Zhiznennyi tsikl nekotorykh kaspiiskikh gammarid. *Dokl. Akad. Nauk SSSR*, **66**, no. 5, 997–999. (The life-cycle of some Caspian gammarids.)

Mordukhai-Boltovskoi, F. D. (1954). Materialy po srednemu vesu vodnykh bespozvonochnykh Dona. Tr. problem i Tematich soveshch. 2. Problemy gidrobiologii vnutrennikh vod. 2. Moscow and Leningrad, 223–241. (Data on the mean weight of aquatic invertebrates of the River Don—2nd Conference on Problems and Topics: Hydrobiological problems of Inland Waters 2.)

Mukerjee, P. (1956). Use of ionic dyes in the analysis of ionic surfactants and other ionic organic compounds. *Analyt. Chem.* **28**, 870–873.

Nauwerck, A. (1963). Die Beziehungen zwischen Zooplankton und Phytoplankton in See Erken. *Symb. bot. upsal.* **17**, 1–164.

Neess, J. C. and Dugdale, R. C. (1959). Computation of production for populations of aquatic midge larvae. *Ecology*, **40**, 425–430.

Nelson, D. J. and Scott, D. C. (1962). Rôle of detritus in the productivity of a rock-outcrop community in a Piedmont stream. *Limnol. Oceanogr.* **7**, 396–413.

Nikolaeva, E. A. (1953). O bikhromatnom metode opredeleniya okislyaemosti organicheskikh veshchtv v prirodnykh vodakh. *Gidrokhim. Mater.* **20**, 68–78. (The bichromate method of determining the oxygen demand of organic substances in natural waters.)

Nikolaeva, E. A. and Skopintsev, B. L. (1961). Bikhromatnaya ikislyaemost' rek ozer Podmoskov'ya i Krupnykh rek Sovetskogo Soyuza. *Gidrokhim. Mater.* **31**, 113–126. (The bichromate oxygen demand of rivers and lakes in the Moscow region and of large rivers of the Soviet Union.)

Novikova, N. S. (1949). O vozmozhnosti opredeleniya sutochnogo ratsiona ryb v estestvennykh usloviyakh (na primere severokaspiiksoi vobly). *Vest. Mosk. gos. Univ. ser. fiz-matem. estestv. nauk.* **5**, 96–112. (The possibility of determining daily fish rations under natural conditions, for example, in the North Caspian roach.)

Novikova, N. S. (1951). Opredelenie sutochnogo ratsiona vobly Severnogo kaspiya neposredstvenno v more. *Vest. Mosk. gos. Univ. ser. fiz-matem. estestv. nauk.* **102**, no. 5, 96–112. (Determining the 24-hour ration of roaches of the North Caspian directly in the sea.)

Obshchie osnovy sovetskoi natsional'noi programmy rabot po izucheniyu produktivnosti presnovodnykh soobshchestv (1966). Moscow and Leningrad. (General principles of the Soviet National Programme of research into the productivity of freshwater aquatic communities.)

Ostapenya, A. P. (1964). Polumikrometody opredeleniya khimicheskogo sostava vodykh organizmov. *Biol. osnov. rybn. kh-va na vnutr. vodoemakh Pribaltiki.* Minsk, 269–272. (Semi-micro methods of determining the chemical composition of aquatic organisms. The biological basis for the fishery industries in Inland Waters of the Baltic region. Minsk.)

Ostapenya, A. P. (1965). Polnota okisleniya organicheskogo veshchestva vodnykh bespozvonochnykh metodom bikhromatnogo okisleniya. *Dokl. Akad. Nauk belorussk. SSR*, **9**, no. 4, 273–276. (The completeness of oxidation of aquatic invertebrate organic matter by the bichromate oxidation method.)

Ostapenya, A. P. (1968). Raschet kaloriinosti sukhogo veshchestva vodnykh organizmov. *Gidrobiol. Zh.*, Kiev, **4**, no. 2, 85–89. (Computing the calorific value of the dry matter of aquatic organisms.)

Ostapenya, A. P. and Kovalevskaya, R. Z. (1965). Soderzhanie vzveshennogo organicheskogo veshchestva v poverkhnostom sloe morskikh vod. *Okean-ologiya*, **5**, no. 4, 649–652. (The content of suspended organic matter in the surface waters of the sea. Trans. C-C.)

Ostapenya, A. P. and Sergeev, A. I. (1963). Kaloriinost' sukhogo veshchestva kormovykh vodnykh bespozvonochnykh. *Vopr. Ikhtiol.* **3**, no. 1, 177–183. (The calorific values of dry matter in aquatic food invertebrates. Trans. FRBC 874.)

Ostapenya, A. P., Pavlyutin, A. P., Babitski, V. A. and Inkina, G. (1968). Transformatsiya energii pishchi nadkotorymi vidami planktonnykh rakoobraznykh. *Zh. obshch. biol.* **28**, no. 3, 334–342. (The transformation of food energy by some species of planktonic crustaceans.)

Paine, R. T. (1964). Ash and caloric determinations of sponge and opisthobranch tissues. *Ecology*, **45**, 384–387.

Paine, R. T. (1966). Endotermy in bomb calorimetry. *Limnol. Oceanogr.* **11**, 126–129.

Peairs, L. M. (1927). Some phases of the relation of temperature to the development of insects. *Bull. W. Va. Univ. agric. Exp. Stn.* **208**, 1–62.

Pechen', G. A. (1965). Produktsiya vetvistousykh rakoobraznykh ozernogo zooplanktona. *Gidrobiol. Zh.* **1**, no. 4, 19–26. (The production of the cladoceran crustaceans in lake zooplankton.)

Pechen', G. A. and Kuznetsova, A. P. (1966). Potreblenie i ispol'zovanie pishchi *Daphnia pulex* (De Geer). *Dokl. Akad. Nauk belorussk. SSR*, **10**, no. 5, 344–347. (Consumption and utilization of food by *Daphnia pulex* (De Geer). Trans. C-C; FRBC 872.)

Pechen', G. A. and Shushkina, E. A. (1964). Produktsiya planktonnykh rakoobraznykh v ozerakh raznogo tipa. *Biol. osnov. rybn. kh-va na vnutr. vodoemakh Pribaltiki*. Minsk, 249–257. (The production of planktonic crustaceans in lakes of diverse types.)

Petipa, T. S. (1965). Pitanie i energeticheskii balans nekotorykh massovykh planktonnykh kopepod Chernogo Morya, otnosyaschikhsya k razlichnym ekologicheskim gruppirovkam. Dokt. diss. (The feeding and energy balance of some dominant planktonic copepods of the Black Sea, belonging to different ecological groupings.)

Petipa, T. S., Pavlova, E. V. and Mironov, G. N. (1966). Energeticheskii balans planktonnykh organizmov iz razlichnykh ekosistem Chernogo Morya. Internat. Oceanogr. Congress, Moscow, 284. (The energy balances of planktonic organisms from different ecosystems of the Black Sea. Trans. Abstracts AEC-TR-6940 (NLL 1769.7F).)

Petrovich, P. G. (1954). Kolichestvennoe razvitie i raspredelenie zooplanktona v ozerakh Zapadnykh oblastei BSSR. *Uchen. Zap. beloruss. gos. Univ. ser. biol.* **17**, 38–71. (The quantitative development and distribution of the zooplankton in lakes of western Belorussia.)

Petrovich, P. G., Shushkina, E. A. and Pechen', G. A. (1964). Raschet produktsii zooplanktona. *Dokl. Akad. Nauk SSSR*, **139**, no. 5, 1235–1238. (Computation of zooplankton production. Trans. C-C.)

Pidgaiko, M. L. (1965). Raschet bio-produktsii nekotorykh vetvistousykh rakoobraznykh. *Vopr. Gidriobiol.* 336–337. (The bio-production calculation of some cladoceran crustaceans.)

Prosyanyi, V. F. and Makina, Z. O. (1960). Pitaniya efektivnosti viroshvannya dvolitkiv koropa pri vysokoushchilnykh posadka'kh. *Navuk. pratsi Ukr. NDU rybn. gosp.* **12**, 19–29. (The feeding efficiency during the rearing of two-year-old carp in high stocking densities.)

Racusen, L. and Johnston, R. (1961). Estimation of protein in cellular material. *Nature*, **191**, 492–493.

Ravera, O. and Tonolli, V. (1956). Body size and number of eggs in diaptomids as related to water renewal in mountain lakes. *Limnol. Oceanogr.* **1**, 118–122.

Raymont, J. E. G. and Conover, R. J. (1961). Further investigations on the carbohydrate content of marine zooplankton. *Limnol Oceanogr.* **6**, 154–164.

Raymont, J. E. G. and Krishnaswamy, S. (1960). Carbohydrates in some marine plankton animals. *J. mar. biol. Ass. U.K.* **39**, 239–248.

Rezvoi, P. D. and Yalynskaya, N. S. (1960). K metodike opredeleniya biomassy planktona i benlosa. *Zool. Zh.* **39**, no. 8, 1250–1252. (Procedures for determining the biomass of plankton and benthos.)

Richman, S. (1958). The transformation of energy by *Daphnia pulex*. *Ecol. Monogr.* **28**, 273–291.

Rubner, M. (1902). "Die Gesetze des Energieverbrauchs bei der Ernährung." Leipzig und Wein.

Shcherbakov, A. P. (1956). Produktivnost' zooplanktona Glubokogo ozera. 1. Rachkovyi zooplankton. *Trudy gidriobiol. Obshch.* **7**, 237–270. (Zooplankton productivity in Lake Glubokoe. 1. The crustacean zooplankton.)

Shorygin, A. A. (1952). "Pitanie i Pishchevye Vzaimootnoshenya ryb Kaspiiskogo Morya." Moscow. ("The Feeding and the Food Relations of Fish in the Caspian Sea.")

Shpet, G. I. and Kozadaeva, T. V. (1963). Energeticheskii i belkovyi balans nagul'nikh karpovych prudov. *Provyshenie ryboproduktibnosti prudov*, **15**, 27–43. (The energy and protein balance of carp-fattening ponds.)

Shushkina, E. A. (1964a). Ratsiony pitaniya populyatsii tsiklopov v ozernom planktone. *Nauch. dokl. vyssh. shkoly. nauki*, **4**, 25–31. (Food rations of cyclopoid populations in lake plankton.)

Shushkina, E. A. (1964b). Razmnozhenie i razvitie planktonnykh tsiklopov pri razlichnykh usloviyakh pitaniya. *Biol. osnovy rybn. kh-va na vnutr. vodoemakh Pribaltiki*, 265–269. (Reproduction and development of plankton cyclopoids under varying trophic conditions.)

Shushkina, E. A. (1964c). Rol' veslonogihk rakoobraznykh v obshchei produktsii zooplanktona ozer. Kand. diss. (The rôle of copepod crustaceans in the general production of lake zooplankton. Trans. FRBC 1051.)

Shushkina, E. A. (1965). Rd' veslonogikh rakoobraznykh v obshchci produktsii zooplanklona ozer. Dok. Diss. Minsk. (The rôle of copepods in the total production of lakes.)

Shushkina, E. A. (1966). Sootnoshenie produktsii i biomassy zooplanktona ozer. *Gidrobiol. Zh.* **2**, no. 1, 27–35. (The ratio of production to biomass in lake zooplankton. *See also* Trans. FRBC 1207.)

Simakov, V. N. (1957). Primenenie fenilantranilovoi kislot pri opredelenii gumusa po metodu I. V. Tyurina. *Pochvovedenie*, **8**, 72–73. (The use of phenylanthranilic acid for humus determinations by Tyurin's method.)

Simakov, V. N. and Tsiplenkov, V. A. (1961). Probirochnyi metod odnovremennogo opredeleniya i okislyaemosti v pochve. *Vest. Leningr. gos. Univ.* **3**, 46–53. (Testing methods of simultaneous determinations of carbon and oxygen demand of soil.)

Sivko, T. N. (1960). K metodike opredeleniya bikhromatnoi okisyaemosti chistykh i zagryaznennykh vod. *Gidrokhim. Mater.* **30,** 190–197. (Procedures for determining the biochromate oxygen demand of pure and polluted waters.)

Skopintsev, B. A. (1947). O kislorodom ekvivalente dlya organicheskogo veshchestva prirodnykh vod. *Dokl. Akad. Nauk SSSR,* **58,** no. 9, 2089–2092. (The oxygen equivalent for organic matter of natural waters.)

Slobodkin, L. B. (1961). Preliminary ideas for a predictive theory of ecology. *Am. Nat.* **95,** 147–153.

Slobodkin, L. B. and Richman, S. (1960). The availability of a miniature bomb calorimeter for ecology. *Ecology,* **41,** 784–785.

Slobodkin, L. B. and Richman, S. (1961). Calories/gm in species of animals. *Nature,* **191,** 299.

Sokolova, N. Yu. (1966). Biologie der Massenarten und Produktivitat den Chironomiden in Utschaustausee. *Veh. Internat. Verein. Limnol.* **16,** 735–740.

Sokolova, N. Yu. (1968). Produktsiya khironomid Uchinskogo vodokhranilishcha —"Metody opredeleniya produktsii vodnykh zhivotnykh". Minsk. (Production of chironomids in the Uchinsk reservoir. *In* "Production of Aquatic Animals", ed. G. G. Winberg. Pt. 2. Trans. FRBC 1177.)

Strickland, J. D. H. (1960). Measuring the production of marine phytoplankton. *Bull. Fish. Res. Bd. Can.* **122,** 1–172.

Strickland, J. D. H. and Parsons, T. R. (1960). A manual of sea water analysis. *Bull. Fish. Res. Bd. Can.* **125,** 1–203.

Stross, R. G., Neess, J. C. and Hasler, A. D. (1961). Turnover time and production of planktonic crustacea in limed and reference portions of a bog lake. *Ecology,* **42,** 237–245.

Sushchenya, L. M. (1961). Nekotorye dannye o kolichestve sestona v vodakh Egeiskogo, Ponicheskogo i Adriaticheskogo morei. *Okeanologiya,* **1,** no. 4, 664–669. (Some data on the seston abundance in the waters of the Aegean, Ionian and Adriatic Seas.)

Sushchenya, L. M. (1964). Kolichestvennye zakonomernosti fil'tratsionnogo pitaniya *Artemis salina* (L.) *Trudy Sevastopol. biol. Sta. Akad. Nauk, USSR,* **15,** 434–445. (Quantitative regularities in the filter feeding of *Artemia salina* (L.).)

Sushchenya, L. M. and Mikhalkovich, V. I. (1961). Kolichestvo sestona v vostochnoi chasti Rizskogo zaliva letom 1958g. *Trudy Nauchno-issled. Inst. ryb. Khoz. Lat. SSR.* **4,** 365–372. (Scston abundance in the eastern part of Riga Bay during the summer 1958.)

Sushchenya, L. M. and Vetrova, S. N. (1957). Vesovaya kharakteristika nekotorykh predstavitelei presnovodnogo zooplanktona. *Uchen. Zap. beloruss. gos. Univ. ser. biol.* **33,** 219–228. (Weight characteristics of some representatives of the freshwater zooplankton.)

Tablitsy khimicheskogo sostava i pitatelnoi tsennosti pischchevykh produktov (1961). Moscow. (Tables of the chemical composition and the nutritive values of foodstuffs.)

Tangl, F. and Farkas, A. (1930a). Beitrage zur Energetik der Ontogenese. 1. Entwicklungsarbeit im Fogelei. *Pfluger's Arch. ges. Physiol.* **93,** 327–376.

Tangl, F. and Farkas, A. (1930b). Beitrage zur Energetik der Ontogenese. III Entwicklungsarbeit im Seidenspinner. *Pfluger's Arch. ges. Physiol.* **98,** 490–546.

Tangl, E. Farkas, A. and Mituchi, M. (1908). V. Entwicklungsarbeit im Huhnerei. *Pfluger's Arch. ges. Physiol.,* **121,** 437–453.

Tauti, M. (1925). *J. imp. Fish. Inst. Tokyo*, **21**, no. 1 (cited in Mednikov, 1965, *Zh. obsh. biol.* **26**, 190).

Taylor, C. C. (1958). Cod growth and temperature (3). *J. Cons. perm. int. Explor. Mer*, **23**, 366–370.

Taylor, C. C. (1960). Temperature, growth and mortality. The Pacific cockle. *J. Cons. perm. int. Explor. Mer*, **26**, 117–124.

Thienemann, A. (1931). Productionsbegrift in der Biologie. *Arch. Hydrobiol.* **22**, 616–621.

Timokhina, A. F. (1964). O produktsii zooplanktona v razlichnykh vodnykh massakh Norvezhskogo morya. *Trudy PINRO*, **16**, 165–181. (Zooplankton production in diverse water masses of the Norwegian Sea.)

Tyurin, I. V. (1934). Novoe vidoizmenenie ob'mnogo metoda opredeleniya gumusa s pomoshch'yu khromovoi kisloty. *Pochvovedenie*, **5-6**, 36–47. (A new modification of the volumetric method of humus determination with chromic acid.)

Ulomski, S. N. (1951). Rol' rakoobraznykh v obshchei biomasse planktona ozer. *Trudy probl. temat. Soveshch. Zool. Inst.* **1**, 121–130. (The rôle of crustaceans in the general biomass of lake plankton.)

Urbakh, V. Yu. (1964). "Biometricheskie Metody." Moscow. ("Biometrical Methods.")

Vasil'eva, G. A. (1959). Issledovanie po ekologii vetvistousykh v svyazi s vyrashchivanien ikh kak zhivogo korma dlya ryb. *Trudy mosk. tekhnol. Inst. ryb. Prom. Khoz.* **10**, 88–138. (Studies of cladoceran ecology to rear them for fish food.)

Vinogradova, Z. A. (1956). K poznanyu khimicheskogo sostave kormovykh organizmov i ryb Chernogo Morya. *Trudy Soveshch. po Fiziol. ryb. Moscow*, 427–436. (Investigating the chemical composition of food organisms and fishes of the Black Sea. Trans. NLL 4271 (8052.98F).)

Vinogradova, Z. A. (1960). Dinamika biokhimichnogo skladu i kaloriinosti planktonu Chernogo Morya v sezonnomu ta geografichnomu aspektakh. *Navuk. zap. Odes'koi biol. sta.* **2**, 3–34. (Changes in the chemical composition and calorific value of Black Sea plankton, seasonally and geographically.)

Vinogradova, Z. A. (1961). Osoblivosti biokhimichnogo skladu ta kaloriinosti fito- i zooplanktonu pivnichnozakhidnoi chastini Chornogo Morya v 1955–1959 pp. *Navuk. zap. Odes'koi biol. sta.* **3**, 3–26. (Special features in the biochemical composition and calorific value of the phyto- and zooplankton in the northwestern Black Sea during 1955–1959.)

Vityuk, D. M. (1964). K metodike opredeleniya syrogo v planktone. *Trudy sevastopol'. biol. Sta.* **17**, 304–308. (Procedures for determining crude fat in plankton.)

Vorob'ev, V. P. (1949). Bentosa Azovskogo Morya. *Trudy azov-chernomorsk. nauch. rybokhoz. Sta.* **13**, 1–129. (The benthos of the Azov Sea.)

Votintsev, K. K. (1955). Vertikal'noe rozpredelenie i sezonnykh variatsii ogranchieskogo veshchestva v vode ozera Baikal. *Dokl. Akad. Nauk SSSR*, **101**, no. 2, 359–363. (The vertical distribution and seasonal variation of organic matter in Lake Baikal water. Trans. C-C.)

V'yushkova, V. P. (1965). O produktivnosti zooplanktona Volgogradskogo vodokhranilishcha. *Trudy saratov. Otd. vses. nauchno-issled. Inst. ozer rech. ryb. Khoz.* **8**, 55–61. (Productivity of zooplankton in the Volgograd reservoir.)

Winberg, G. G. (1954) Nekotorye kolichestvennye dannye po biomass planktona ozer BSSR. *Uchen. Zap. beloruss. gos. Univ. ser. biol.* **17**, 20–37. (Some quantitative data on planktonic biomass in Belorussian lakes.)

Winberg, G. G. (1956). "Intensivnost' Obmena i Pishchevye Potrebnosti Ryb." Minsk, 253 pp. (The Metabolic Intensity and Food Requirements of Fish. Trans. FRBC 194. *See also* FRBC 362.)

Winberg, G. G. (1962). Energeticheskii printsip izucheniya troficheskikh svyazei i produktivnosti ekologicheskikh sistem. *Zool. Zh.* **41**, no. 11, 1618–1630. (The energy principle in studying food relations and productivity of ecological systems. Trans. FRBC 433.)

Winberg, G. G. (1964). Puti kolichestvennogo izucheniya potrebleniya i usvoeniya pishchi vodnym zhivotnymi. *Zh. obshch. Biol.* **25**, no. 4, 254–266. (Pathways of quantitative study of food consumption and assimilation by aquatic animals.)

Winberg, G. G. (1966). Skorost' rosta i intensivnost' obmena u zhivotnykh. *Usp. sovrem. Biol.* **61**, 274–293. (Growth rates and metabolic intensity of animals. Trans. C-C.)

Winberg, G. G. (1967). Osnovn'e napravleniya v izuchenii biologicheskogo balansa ozer. *In* "Krugovovot Veshchestva i Enrgii v Ozernykh Vodoemakh," 132–147. Nauka, Moscow. (Basic concepts in the biotic balance of lakes. *In* "Circulation of Matter and Energy in Lake Reservoirs".)

Winberg, G. G. and Anisimov, S. I. (1966). Matematicheskaya model' vodnoi ekosistemy. *In* "Fotosinteziruyushche Sistemy Vysokoi Produktivnosti." Moscow, 213–223. (A mathematical model of an aquatic ecosystem. *In* "Photosynthetic Systems of High Productivity." *See also* Trans. FRBC 1571.)

Winberg, G. G. and Koblents-Mishke, O. I. (1966). Problemy pervichnoi produktsii vodoemov. *Ekologiya vodnykh organizmov. Akad. Nauk SSSR, Moscow,* 50–62. (Questions on primary production in waters. Trans. FRBC 1043.)

Winberg, G. G. and Lyakhnovich, V. P. (1965). "Udobrenie Prudov." Moscow. 271 pp. ("The Fertilization of Ponds." Trans. FRBC 1339.)

Winberg, G. G. and Platova, T. P. (1951). Biomassa planktona i rastvorennoe organicheskoe veshchestvo v vode ozer. *Byull. MOIP, otd. biol.* **56**, no. 2, 24–37. (Planktonic biomass and dissolved organic matter in lake water.)

Winberg, G. G. and Zakharenkov, I. S. (1950). K kolichestvennoi kharakteristike roli planktona v krugovorote veshchestv v ozerakh. *Dokl. Akad. Nauk. SSSR.* **73**, no. 6, 1037–1039. (A quantitative characterization of the part played by plankton in the turnover of matter in lakes.)

Winberg, G. G., Ivlev, V. S., Platova, T. P. and Rossolimo, L. L. (1934). K metodike opredelennya organicheskogo veshchestva. Opyt kaloricheskoi otsenki kormovykh zapasov vodoema. *Trudy limnol. Sta. Kosino,* **18**, 25–40. (Procedures for determining organic matter. Experiment on a calorific evaluation of food supplies in a body of water.)

Winberg, G. G., Pechen', G. A. and Shushkina, E. A. (1965). Produktsiya planktonnych rakoobraznykh trekh ozerakh raznogo tipa. *Zool. Zh.* **44**, 676–688. (The production of planktonic crustaceans in three different types of lake. Trans. NLL RTS 6019.)

Wright, J. C. (1965). The population dynamics and production of *Daphnia* in Canyon Ferry Reservoir, Montana. *Limnol. Oceanogr.* **10**, 583–591.

Yablonskaya, E. A. (1947). Opredelenie produktsii lichinok *Chironomus plumosus* Medvezk'ikh ozer. Kand. Diss. (Determining the production of *Chironomus plumosus* larvae in the Medvezh'e Lakes.)

Yablonskaya, E. A. and Lukonina, N. K. (1962). K voprosu o produktivnosti Aral'skogo Morya. 1. Intensivnost' obrazovaniya produktsii zooplanktona. *Okeanologiya*, **2**, 2. (On the question of the productivity of the Aral Sea. 1. Intensity of development of zooplankton production.)

Yaroshenko, M. F. and Naberezhnyi, A. I. (1955). O biologicheskoi produktivnosti kormovoi gidrogauny v prudakh dlya karpov. *Izv. Moldavsk. fil. Akad. Nauk SSSR*, **6**, no. 26, 117–138. (The biological productivity of food hydrofauna in carp ponds.)

Yashnov, V. A. (1940). "Planticheskaya Produktivnost' Severnykh Morei SSSR." Moscow. (Plankton Productivity in the Northern Seas of the USSR.")

Zhukova, N. A. (1953). Tsiklomorfoz u dafnii. *Uch. zap. Leningr. ped. in-ta*, **3**, 85–148. (Cyclomorphosis in daphnia.)

Some relevant translations: FRBC 1197, 1224, 1225, 1226, 1228.

G

AUTHOR INDEX

163

SUBJECT INDEX

A

A = assimilation, 10
Absolute growth increment, 35
Acanthodiaptomus denticornis,
  biomass and production, 96
  rate of embryonic development, 50, 67
  renewal rates, 97
Acartia clausi, 58
  growth, 116, 122
  K_2, 118
  metabolism, 118
  production, 115, 120, 121
Acclimation, 59
Accuracy,
  of population density determinations, 73
  of weighing, 15
Adaptation to temperature, 58, 59
Age classes, 66-67
  frequency distribution, 106
  structure, 6
    determination in a population, 88
    and growth, 102, 105
Allometric growth equation, 37
Amphipod production, 67-69
  see also Species
Anatopynia gr. *varia,*
  head capsules and instars, 79
Angara, River, USSR, 67
Anodonta anatina,
  length–weight relationship, 34, 35
Anodonta piscinalis,
  ash content of, 15
Anuraeopsis fissa, growth increments, 125
Aral Sea, 86
Arba, 67
Arcthodiaptomus bacillifer,
  biomass and production, 96
Arcthodiaptomus spinosus var. *fadeevi,*
  biomass and production, 96
Arrhenius, van't Hoff formula, 51
Artemia salina, K_2, 118
Ascomorpha sp., sampling, 124
Asellus aquaticus,
  length–weight relationship, 34
Ash content, determination, 11, 15-17, 20
  of freshwater molluscs, 15, 17

potassium content of, 16
Asplanchna priodonta,
  biomass, 125
  growth increments, 125
  production, 125
  P/B, 125
Assimilability, 10
Assimilation, 6, 7, 9
Azov Sea, 131, 133

B

Baikal, Lake, USSR, 69, 108, 110
Barents Sea, 71, 83
Base Line Lake, USA, 99
Batorin, Lake, BSSR, 117, 137
Beloye, Lake, USSR, 77
Benthic organisms,
  homotopic animals, 67
  production estimation from fish con-
    sumption, 131-143
  production estimations, 65-83
Bering Sea, 85
Bichromate oxidation, 21-25
Biomass, 3, 7-9
  determination of, 11-17
    for benthos, 66
  in samples, 88
  initial biomass, 65
  of age groups, 107
  of chironomid larvae, 73
  of rotifers, 125
  residual biomass, 11, 65
Birth rate, 99
Biuret reaction, 28
Black Sea, 115
Bol'shoe, Lake, USSR, 74, 77
Bomb calorimeter, 16, 19, 27
  microcalorimeters, 20
Bosmina sp., length–weight relationship, 34
  as food for fish, 140
Boysen-Jensen method, 65-67, 82, 85
  cf. other methods, 78
Brachionus calyciflorus,
  food and metabolism, 127
  K_2, 118

167